CHEM3

NCEA LEVEL 3

DEBORAH HAY

NELSON
A Cengage Company

Chem3 NCEA Level 3
1st Edition
Deborah Hay

Cover design: Cheryl Smith, Macarn Design
Text designer: Cheryl Smith, Macarn Design
Production controller: Siew Han Ong

Acknowledgements
Photograph on page 121 and back cover courtesy of Fundamental Photographs.

National Library of New Zealand Cataloguing-in-Publication Data
A catalogue record for this book is available from the National Library of New Zealand.

ISBN 978 0 17 035261 1

Cengage Learning Australia
Level 7, 80 Dorcas Street
South Melbourne, Victoria, Australia 3205

Cengage Learning New Zealand
Unit 4B Rosedale Office Park
331 Rosedale Road, Albany, North Shore 0632, NZ

For learning solutions, visit **cengage.com.au**

Printed in China by 1010 Printing International Limited
3 4 5 6 7 8 23

contents

1 3.4 Demonstrate understanding of thermochemical principles and the properties of particles and substances (91390) 4

Learning outcomes 4
Pre-test — What do you know? 5
Recapping Level 2 6
Electron arrangement 7
Trends of the periodic table 10
Lewis structures and shapes of molecules 17
Forces of attraction between particles 24
Specific heat capacity 30
Enthalpy, ΔH 32
Entropy 34
Hess's Law 35

Exam-type questions 41

2 3.5 Demonstrate understanding of the properties of organic compounds (91391) 51

Learning outcomes 51
Pre-test — What do you know? 52
Recapping Level 2 53
Isomerism 53
Hydrocarbons 56
Alcohols 59
Haloalkanes 67
Amines 69
Aldehydes and ketones 72
Carboxylic acids and acid chlorides 77
Esters and amides 80
Polymers 84
Proteins 85
Organic reaction scheme summary 91

Exam-type questions 92

3 3.6 Demonstrate understanding of equilibrium principles in aqueous systems (91392) 102

Learning outcomes 102
Pre-test — What do you know? 103
Recapping Level 2 105
Solubility equilibria 108
Solubility, s 109
The common ion effect 113
Predicting precipitation 119
Complex ions and solubility 121
Species in solution 125
pH and conductivity of acids 130
pH and conductivity of bases 132
Buffers 135
Titration curves 138

Exam-type questions 148

Internal Standards review

3.1 — Carry out an investigation in chemistry involving quantitative analysis (91387) 159
3.2 — Demonstrate understanding of spectroscopic data in chemistry (91388) 161
3.3 — Demonstrate understanding of chemical processes in the world around us (91389) ... 164
3.7 — Demonstrate understanding of oxidation-reduction processes (91393) 165

Practice assessment section

91390 Demonstrate understanding of thermochemical principles and the properties of particles and substances 168
91391 Demonstrate understanding of the properties of organic compounds 175
91392 Demonstrate understanding of equilibrium principles in aqueous systems 180

Glossary of terms 185
Periodic table 187
Answers 188

CHAPTER ONE

3.4 Demonstrate understanding of thermochemical principles and the properties of particles and substances (91390)

Learning outcomes

Tick off when you have studied these ideas in class and when you have revised that section prior to your assessment.

	Learning outcomes	In class	Revision
1	Write electron configurations using s, p, d, f notation for the first 36 atoms and ions.		
2	Describe the trend in electronegativity in the periodic table.		
3	Describe the trend in ionisation energy in the periodic table.		
4	Describe the trends in atomic and ionic radii size in the periodic table.		
5	Draw Lewis structures and explain the shapes of molecules/polyatomic ions with up to six atoms in their structure.		
6	Explain the attractive forces and properties present in an ionic compound.		
7	Explain the attractive forces and properties present in a metallic solid.		
8	Explain the attractive forces and properties in molecular solids.		
9	Explain the polarity of molecular solids.		
10	Explain hydrogen bonding.		
11	Explain and calculate enthalpy changes in chemical reactions.		
12	Calculate specific heat capacity of different substances.		
13	Recognise and be able to calculate the following: $\Delta_c H°$, $\Delta_f H°$, $\Delta_r H°$, $\Delta_{vap} H°$, $\Delta_{sub} H°$ and $\Delta_{fus} H°$.		
14	Explain entropy changes in chemical reactions.		
15	Use Hess's Law to calculate enthalpy changes of reaction.		

ISBN: 9780170352611

Pre-test — What do you know?

1 Write the electron arrangement for the following atoms and ions:

a magnesium, Mg ______

b chlorine, Cl ______

c magnesium ion, Mg^{2+} ______

d chloride ion, Cl^- ______

2 What happens to the electronegativity of an element as you move across the periodic table?

______

3 What happens to the electronegativity of an element as you move down a group?

______

4 What change occurs across the periodic table as ions form?

______

______

5 Draw the Lewis structure for the following compounds:

a NH_3, ammonia

b BF_3, boron trifluoride

6 What are the forces of attraction between the following particles?

a Magnesium and chloride ions ______

b Magnesium atoms ______

c Ammonia molecules ______

7 What is enthalpy a measure of and what letter stands to represent it?

______

______

8 If you combust 30 grams of methane, CH_4, in a plentiful supply of oxygen and it releases 200 kJ of energy, what is the change in enthalpy for this reaction? (The equation is written below.)

$CH_{4\,(g)} + 2O_{2\,(g)} \longrightarrow CO_{2\,(g)} + 2H_2O_{(g)}$ $M(CH_4) = 16.0\ gmol^{-1}$

______

______

______

Recapping Level 2

In Level 2 there are two main sections to the topic that you will need to draw upon as you complete this standard at Level 3.

Bonding and properties

There are four main types of solid that you needed to be able to explain in Level 2 in terms of their bonding and structure.

Property	Ionic	Metallic	Molecular	Covalent network
Structure	Ionic solids are made up of ions that are attracted to each other because they are oppositely charged (an electrostatic attraction).	Metallic solids are made up of metal nuclei forming an electrostatic attraction with their delocalised valence electrons.	Molecular solids are made up of molecules that can be either polar or non-polar. There are covalent bonds present within the molecules and intermolecular bonds between the molecules.	Covalent network solids are made up of atoms that are bonded by covalent bonds where electrons are shared.
Solubility	They are generally very soluble in polar solvents like water, as the attraction between the water molecules and the ions is strong enough to overcome the ionic bonds.	They are insoluble in both polar and non-polar solvents, as the force of attraction between the solvents and the metal atoms is not strong enough to overcome the metallic bonds.	Polar molecules will dissolve in polar solvents and non-polar molecules will dissolve in non-polar solvents, as the force of attraction between the molecules and the solvent is strong enough to overcome the intermolecular bonds.	They are insoluble in both types of solvent.
Melting points/ boiling points	They have high melting and boiling points due to the large amount of energy required to overcome the ionic bonds.	They have high melting and boiling points due to the large amount of energy required to overcome the metallic bonds.	They have lower melting and boiling points generally. However, polar solids generally have higher ones than non-polar solids.	They have very high melting and boiling points due to the very strong covalent bonds.
Conductivity	They are conductive when molten or aqueous as their ions are free to move to conduct a current.	They will conduct a current, as their electrons are delocalised.	They cannot conduct, as they have no free-moving charged particles.	They are generally non-conductive. However, graphite will conduct, as each carbon atom in it has a free electron.

ISBN: 9780170352611

Property	Ionic	Metallic	Molecular	Covalent network
Other special properties	They are brittle, as when struck the ion's regular lattice arrangement is disrupted causing like charged ions next to each other breaking up the ionic bonds.	They are malleable, which means they can be bent, and ductile, which means they can be drawn into wires. Both of these occur because when bent, the metallic bonds are not broken.	None.	They are extremely hard.
Examples	NaCl, $MgBr_2$, Al_2O_3	Mg, Be, Na	Cl_2, O_2, H_2	Graphite, diamond, SiO_2

Note:
Electronegativity — the ability of an atom to attract a pair of bonding electrons.
Polar — a covalent bond where electrons are shared unequally between two atoms due to their difference in electronegativity.
Non-polar — a covalent bond where the electrons are shared equally between two atoms, as there is no difference in electronegativity between the two atoms.

Enthalpy

Enthalpy is a measure of the heat content of a substance per mole and has the symbol H. If we know the amount of energy released or absorbed from a given reaction and the moles of a reaction, we can calculate the change in enthalpy, ΔH, for this reaction. An example is shown below where we want to calculate the amount of heat released:

Example: Making methane

$$C + 2H_2 \longrightarrow CH_4 \qquad \Delta H = -17.9 \text{ kJmol}^{-1}$$

If we are given 2 mol of carbon, then the total heat released will be:
Moles (n) x change in enthalpy, ΔH = 2 mol x 17.9 $kJmol^{-1}$ = 35.8 kJ released

An **exothermic** reaction is a reaction where heat is released into the surroundings and so it has a negative change in enthalpy value. Making bonds is an exothermic process.

An **endothermic** reaction is a reaction where heat is absorbed from the surroundings and so it has positive change in enthalpy value. Breaking bonds is an endothermic process.

Electron arrangement

At Level 2, the electron arrangement of an atom only describes how many electrons there may be found at a certain level of energy from the nucleus of an atom, but it is not specific. At Level 3, we want to give an electron its address, so we need to know a bit more about where electrons exist and the probabilities of finding them in a certain area of space.

As in Level 2, the electrons will start filling as close to the nucleus as possible before filling at greater distances from the nucleus because this is the lowest energy configuration.

The energy levels are the first part of the address of an electron. They determine the amount of energy or quanta in order for an electron to move to this level; 1 is the first level, then 2, and so on.

Next comes the orbitals; these are areas of probability where you may find any electron at a particular quantum level. These orbitals are called s, p, d and f; each has its own unique shape, as shown on the following page.

ISBN: 9780170352611

Shapes of the electron orbitals

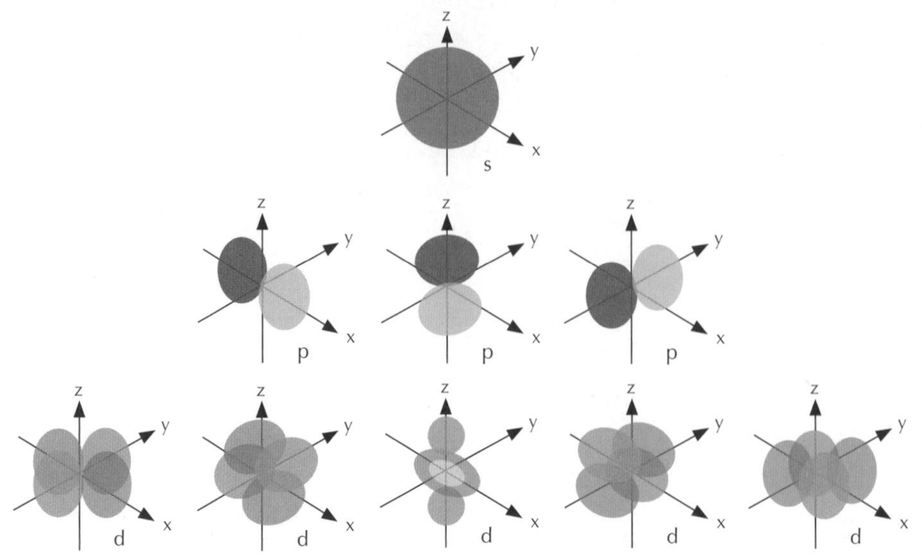

In the first quantum level, 1, there is only an s orbital; in the second, there is an s and a p; in the third, there is an s, p and d; and in the fourth, there is an s, p, d and f, as there is progressively more space as the electrons are further from the nucleus.

Each suborbital holds a maximum of two electrons, which are spin paired (that is, spinning in opposite directions). You can see from the diagram above, though, that the s orbital has no suborbitals and so will hold only two electrons; the p orbital has three suborbitals so in total holds six electrons; the d has five suborbitals so can hold 10 electrons; and the f has seven suborbitals and so can hold up to 14 electrons.

For example, the electron arrangement for hydrogen is $1s^1$ — the superscript 1 next to the s tells you how many electrons there are.

The exact order for filling up the electron orbitals is as follows: 1s2s2p3s3p4s3d4p4d. (Note we fill up the 4s before the 3d since it is lower energy despite being further out from the nucleus.) Electrons will also half-fill all suborbitals before pairing up, as this creates less repulsion.

There are two exceptions to this rule because they have lower energy states by having a half-filled or fully filled 3d orbital. They are Cr, which has the arrangement $1s^2 2s^2 2p^6 3s^2 3p^6 3d^5 4s^1$, and Cu, $1s^2 2s^2 2p^6 3s^2 3p^6 3d^{10} 4s^1$. We can also write these in shorthand using the noble gases, so Cr can also be written $[Ar]3d^5 4s^1$ and Cu, $[Ar]3d^{10} 4s^1$.

1.1: Atom electron arrangement

Write the electron arrangement in s, p, d, f notation for the following elements.

a Boron, B ____________________

b Magnesium, Mg ____________________

c Chlorine, Cl ____________________

d Scandium, Sc ____________________

e Iron, Fe ____________________

f Zinc, Zn ____________________

g Chromium, Cr ____________________

You may also be asked to write the electron arrangement for an ion. Note that the 4s orbital is emptied before the 3d.

ISBN: 9780170352611

1.2: Ion electron arrangement

Write the electron arrangement in s, p, d, f notation for the following ions.

a Chloride ion, Cl^- ____________________

b Oxide ion, O^{2-} ____________________

c Magnesium ion, Mg^{2+} ____________________

d Iron (II) ion, Fe^{2+} ____________________

e Iron (III) ion, Fe^{3+} ____________________

f Zinc ion, Zn^{2+} ____________________

g Copper (II) ion, Cu^{2+} ____________________

We can also fill in an orbital diagram to show where electrons may be found.

For example, the orbital diagram for aluminium is shown below.

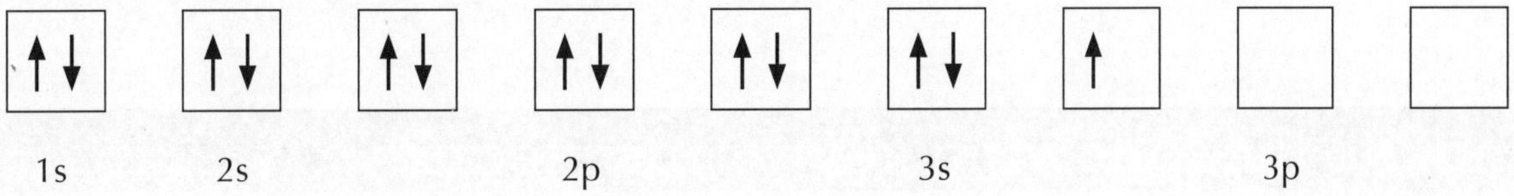

Remember, electrons will half-fill suborbitals if they can, before fully filling them.

1.3: Orbital diagrams

Fill in the orbital diagrams below.

a Hydrogen, H

1s 2s 2p 3s 3p 3d 4s

b Lithium, Li

1s 2s 2p 3s 3p 3d 4s

c Lithium ion, Li^+

1s 2s 2p 3s 3p 3d 4s

d Aluminium ion, Al^{3+}

1s 2s 2p 3s 3p 3d 4s

e Scandium, Sc

1s 2s 2p 3s 3p 3d 4s

ISBN: 9780170352611

f Chromium, Cr (don't forget it is an exception)

1s	2s		2p		3s		3p				3d			4s

g Copper, Cu (don't forget it is an exception)

1s	2s		2p		3s		3p				3d			4s

h Copper (II) ions, Cu^{2+}

1s	2s		2p		3s		3p				3d			4s

What do the following key words mean?

Electron	
Orbital	
Quantum level	

FACT

Wolfgang Pauli came up with the theory that no electron can exist in the same place as another electron, hence even though electrons are paired up in suborbitals, they will spin in opposite directions.

Pauli was born in 1900 in Austria and died in 1958. He was referred to in the physics community as the 'conscience of physics' because he was so critical of any papers that were published. He was also known for the Pauli effect because there was a high chance that equipment would break whenever he walked into a lab! In 1945, he won the Nobel Prize in Physics after being nominated by Albert Einstein.

Trends of the periodic table

There are three trends that you need to know about and be able to explain about the periodic table. They are:

1 Electronegativity
2 Ionisation energy
3 The size of the atomic and ionic radii.

1 Electronegativity

This is the ability of an atom to attract a bonding pair of electrons towards itself.

ISBN: 9780170352611

Across the periodic table electronegativity ***increases*** as the electrons are being added to the same quantum level and so are no further from the nucleus (relatively), but there are more protons being added into the nucleus so there is an increase in effective nuclear charge (the attraction between the protons in the nucleus and the valence electrons). This means there is more effective nuclear charge to also pull on the electrons in a bond too.

Down a group in the periodic table electronegativity ***decreases*** as the valence electrons are shielded from the nuclear charge because of the extra quantum level that is added at each period.

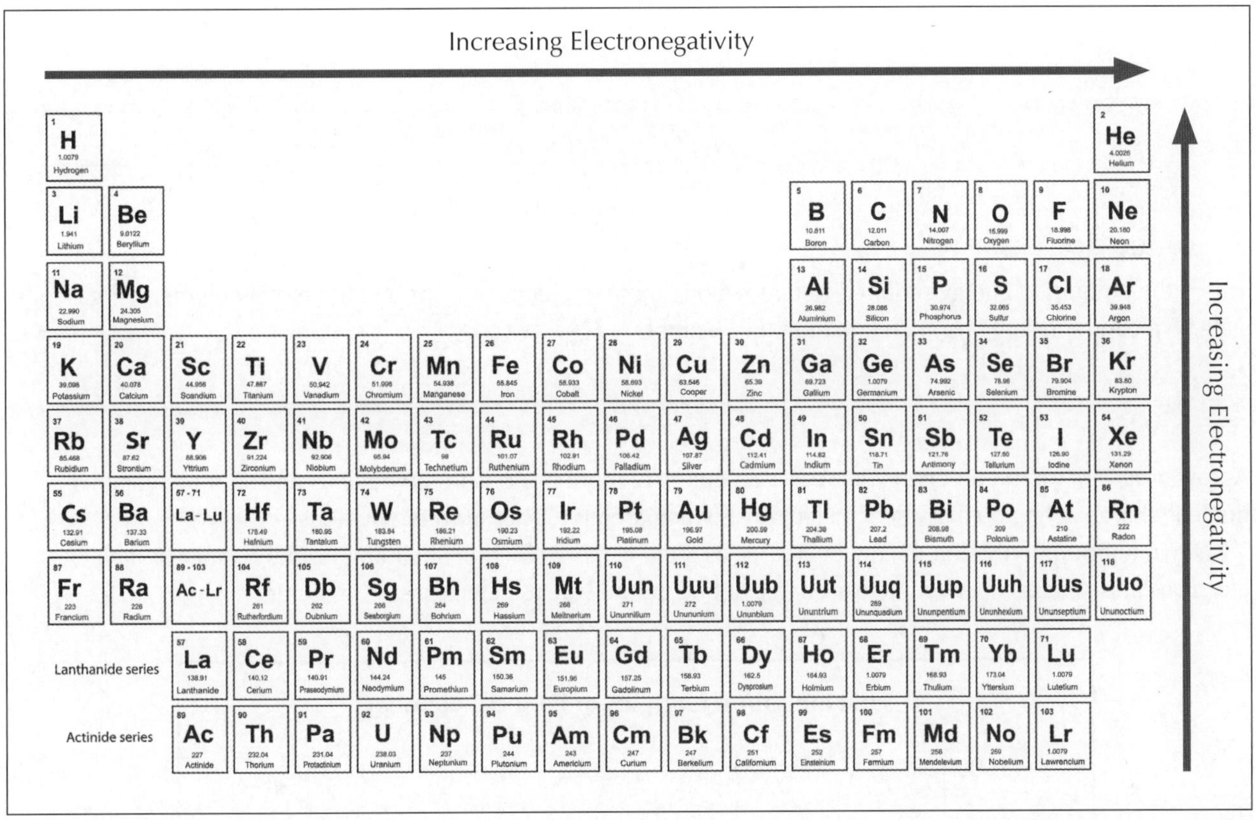

1.4: Electronegativity

a What is the definition for electronegativity?

b Circle the element which has a higher electronegativity in the following pairs and explain why this is.

i Li or B

ii Li or Na

iii F or Br

2 Ionisation energy

This is the amount of energy required to remove one mole of electrons from one mole of an element in gaseous state. We can represent this in an equation, for example:

$$F_{(g)} \longrightarrow F^{+}_{(g)} + e^{-}$$

Across a period, ionisation energy increases as there are more protons in the nucleus and so more effective nuclear charge acting on the valence electrons, which are no further from the nucleus.

Down a group, ionisation energy decreases as there is more electron shielding from the inner orbitals of electrons and so less effective nuclear charge acting on the valence electrons.

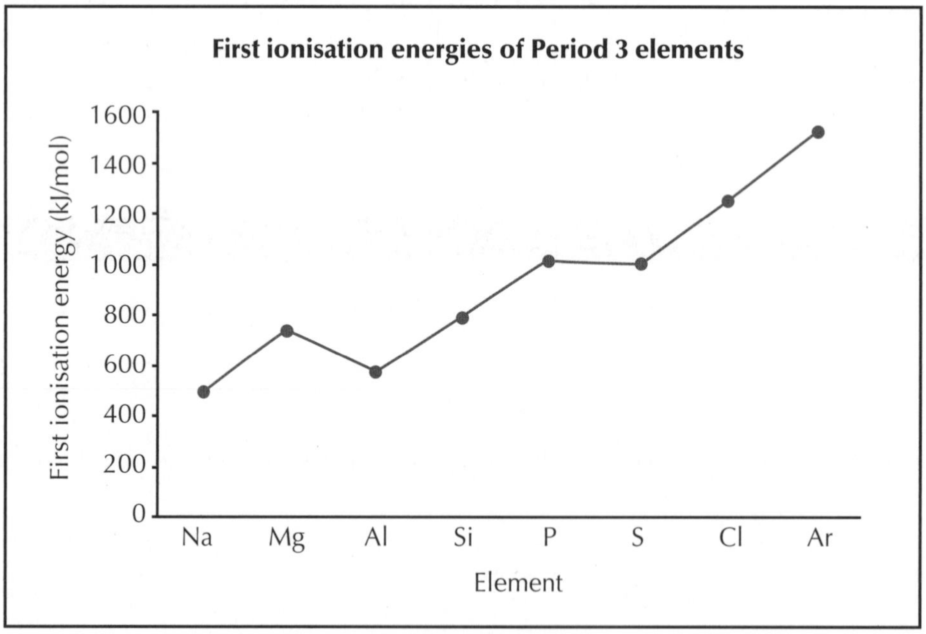

You will see in any graph of ionisation energy that it is not a linear relationship. This is because there is a little extra stability by having a completely filled orbital or half-filled orbital. For example, magnesium has the electron arrangement $1s^2 2s^2 2p^6 3s^2$ and so it is a little more stable than aluminium, which has the electron arrangement $1s^2 2s^2 2p^6 3s^2 3p^1$.

ISBN: 9780170352611

1.5: Ionisation energy

a What is the definition for ionisation energy?

b Write the first ionisation energy equation for beryllium, Be.

c Circle the element which has a higher first ionisation energy in the following pairs and explain why this is.

i Li or B

ii Li or Na

iii F or Br

d Explain why phosphorus has a slightly higher first ionisation energy than sulfur.

ISBN: 9780170352611

3 The size of the atomic and ionic radii

The atomic radius is the distance between two atomic nuclei divided by two; the ionic radius is the distance between two ionic nuclei divided by two.

Across the periodic table, atomic radii ***decrease*** in size as there is more effective nuclear charge due to the increase in the number of protons in the nucleus with the valence electrons being added to the same quantum level.

Down the periodic table, the atomic radii ***increases*** as there is more electron shielding as another whole quantum level is added on in each period.

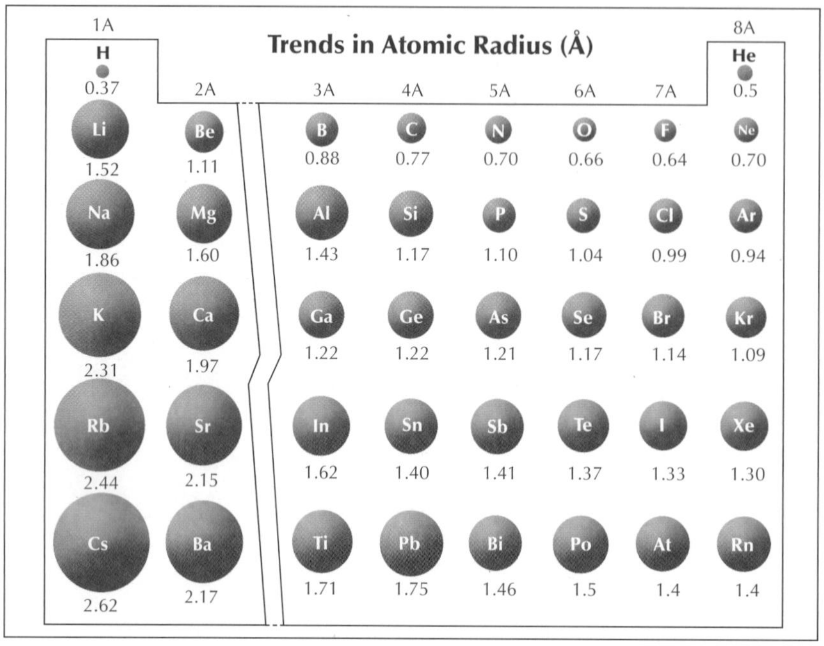

Across the periodic table, ionic radii, on the other hand, ***increase*** as electrons go from being taken away on the left-hand side, increasing the number of protons to electrons and decreasing the distance between the nucleus and the valence electrons, to electrons being added on, decreasing the number of protons to electrons while increasing the repulsion between the valence electrons.

Down the periodic table, ionic radii still ***increase*** in size as there is still one more quantum level being added at each period.

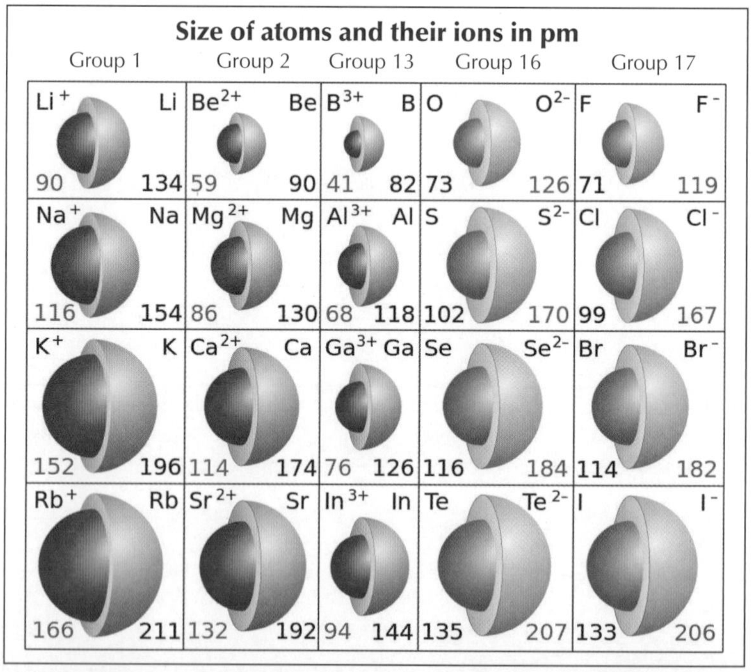

EXAMINER'S TIP

The loss of an energy level of an atom when becoming an ion will have more of an effect on the size than the fewer protons found in a nucleus.

For example, if we compare the size of Na^+ to Cl^-, Na^+ will be much smaller than Cl^- not only because it has more protons to electrons but more importantly it has one less energy level.

ISBN: 9780170352611

1.6: Atomic and ionic radii

a What is an atomic radius?

b Circle the element which has a larger atomic radius in the following pairs and explain why this is.

i Li or B

ii Li or Na

iii F or Br

c Circle the ion which has a larger ionic radius in the following pairs and explain why this is.

i F^- or Cl^-

ii Na^+ or Cl^-

ISBN: 9780170352611

What do the following key words mean?

Electronegativity	
Effective nuclear charge	
Ionisation energy	
Atomic radius	
Ionic radius	

CHECKPOINT 1
WHAT HAVE YOU LEARNED SO FAR?

Electron arrangements and trends of the periodic table

1 Write the electron arrangements in s, p, d, f notation for the following:

a Ca ______________________

b V ______________________

c P ______________________

d Si ______________________

e Cu^{+} ______________________

f Be^{2+} ______________________

g S^{2-} ______________________

h Co^{2+} ______________________

2 Explain why O has a higher electronegativity value than S.

____

____

____

____

____

3 Explain why helium has a higher first ionisation energy than neon.

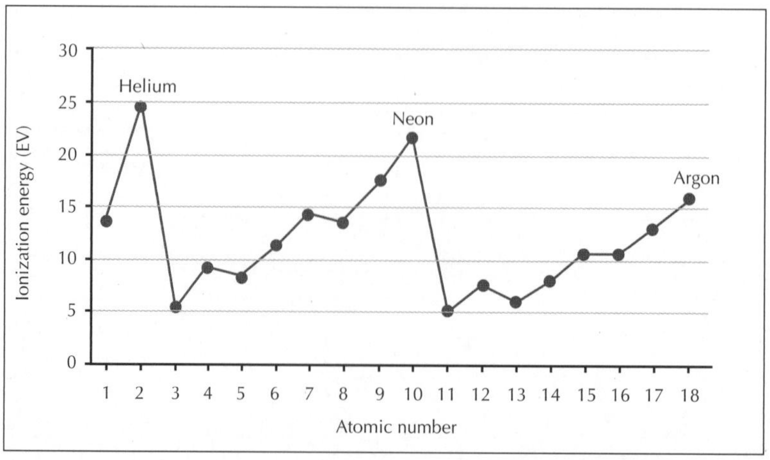

 ISBN: 9780170352611

CHECKPOINT 1
WHAT HAVE YOU LEARNED SO FAR?

4 Explain why the oxide ion is so much larger than the oxygen atom.

Lewis structures and shapes of molecules

Lewis structure

These structures show the placement of bonds and lone pairs of electrons in molecules and polyatomic ions. This year you will need to be able to draw them for molecules with expanded octets as well as for polyatomic ions in order to be able to describe their shape.

Elements beyond silicon in the periodic table can have an expanded octet of electrons surrounding the central atom. When you draw your Lewis structure for one of these molecules, make sure all atoms have eight electrons surrounding them and place any leftover electrons or atoms around the central atom.

Example: PCl_5 – this molecule has 5 electrons from P and 7 x 5 electrons from the 5 x Cl = 40 electrons in total.

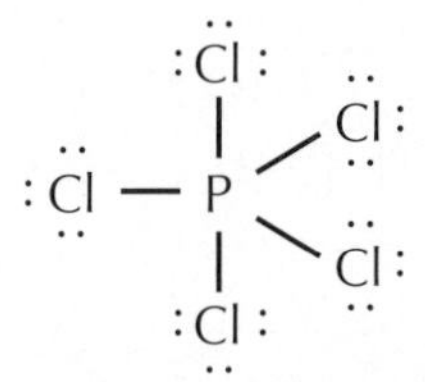

As there is only one P atom, this must be the central atom. It is also the least electronegative atom.

1.7: Expanded octet Lewis structures

Draw the Lewis structures of the following molecules.

a SF_6

b XeF_4

c ICl_3

d SF_4

Polyatomic ions

When drawing the Lewis structure for an ion, you must add on or take away the charge of the ion (for example, if it is a -1 ion, then add on one electron, and if it is a +2 ion, take away two electrons) and you must draw square brackets about the whole structure, placing the charge outside.

Example: H_3O^+

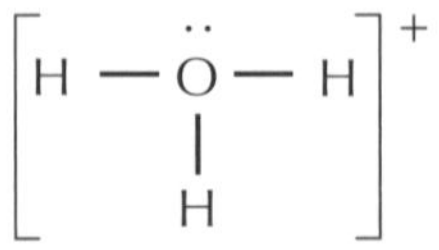

Note the rest of the rules for drawing Lewis structures remain the same.

1.8: Polyatomic ion Lewis structures

Draw the Lewis structures of the following polyatomic ions.

a CO_3^{2-}

b NH_4^+

c SO_3^{2-}

d PCl_6^-

Shapes of molecules

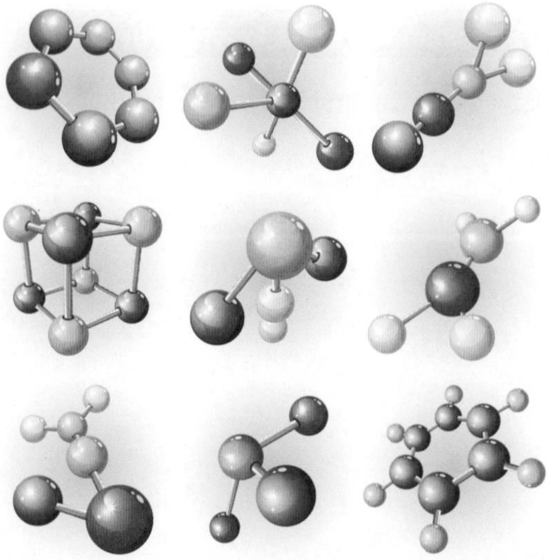

There are many more shapes you need to be able to predict and draw at Level 3, which are shown below. Remember you must always draw a Lewis structure in order to help predict the shape and use VSEPR (valence shell electron pair repulsion) theory to help explain the shape.

There are two main points to VSEPR theory to remember:

1. Electrons repel each other to be as far apart as possible.
2. Lone pairs of electrons repel more strongly than bonded pairs of electrons.

ISBN: 9780170352611

Name of shape	Example	Bond angles
Linear	CO_2 O = C = O	180°
Bent	H_2O or SO_2 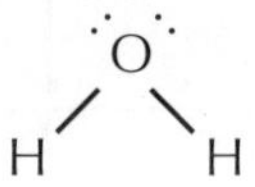 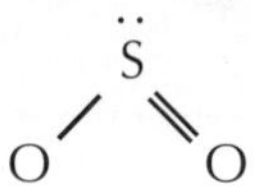	< 109° (with two lone pairs) < 120° (with one lone pair)
Trigonal planar	BF_3 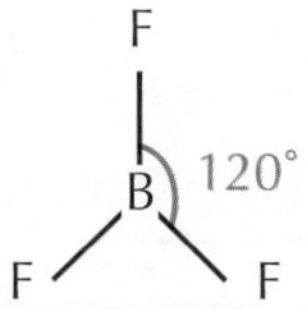	120°
Trigonal pyramid	NH_3 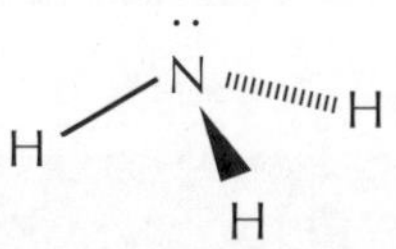	< 109°
T-shaped	ICl_3 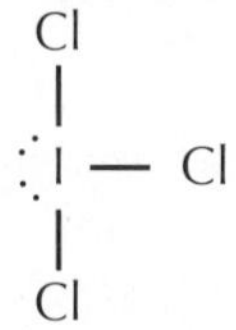	90° (between bonded atoms)
Tetrahedral	CH_4 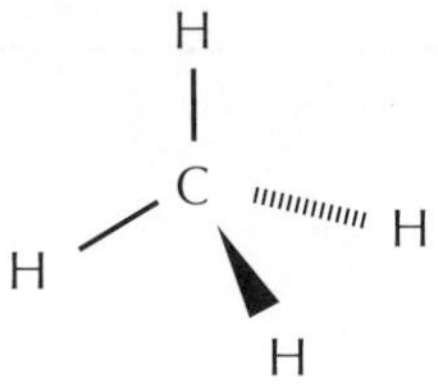	109°
Seesaw or distorted tetrahedron	SF_4 	90° and 120°
Square planar	XeF_4 	90°
Trigonal bipyramid	PCl_5 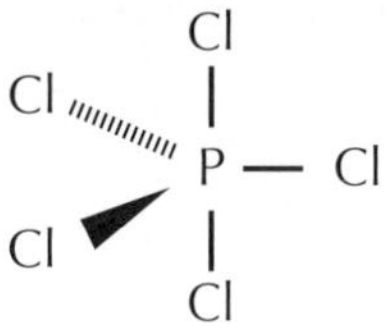	90° and 120°

Name of shape	Example	Bond angles
Square pyramid	ICl_5 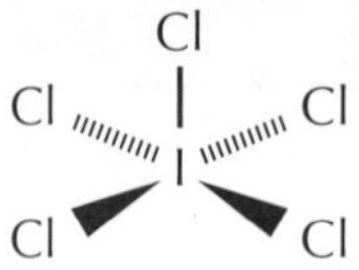	90°
Octahedral	SF_6 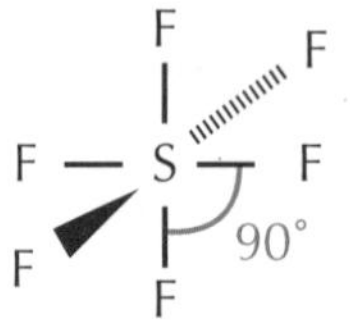	90°

1.9: Shapes of molecules

Draw the Lewis structure, predict the shape and draw the shape diagram for the following molecules.

Molecule	Lewis structure	Shape
CF_4		
NF_3		
SF_3^+		
PCl_6^-		

ISBN: 9780170352611

Molecule	Lewis structure	Shape
PCl_4^-		
BrF_3		
I_3^-		
PH_3		

EXAMINER'S TIP

When explaining the polarity of a particular shape, make sure you talk about whether the molecule is symmetrical or not.

For example, SF_6, which is octahedral, is symmetrical, but SF_4, which is seesaw, is not.

When explaining the shape of a molecule, you must talk about the repulsion between bonding and non-bonding electron pairs.

For example, in BrF_3 there are two non-bonding pairs and three bonding pairs about the central Br atom, which repel each other to be as far apart as possible, with the lone pairs repelling just slightly more strongly.

Note: See Recapping Level 2 section for a refresher on polarity and polar bonds.

INVESTIGATION 1

Shape and polarity

AIM: To build models of different shapes and explain why the shapes occur or whether or not they are polar or non-polar.

EQUIPMENT:
Molymods or plasticine and toothpicks.

TASK:

1 Build a model of the ClF_3 molecule and draw the 3D structure in the box below.

Explain the shape that this molecule has using VSEPR theory. In your answer you must include:

- the shape
- the number of lone and bonding pairs about the central atom
- the repulsion between those electron pairs
- why the molecule has the particular shape that it does and no other.

2 Build a model of the AsF_5 molecule and draw the 3D structure in the box below.

Explain the shape that this molecule has using VSEPR theory. In your answer you must include:

- the shape
- the number of lone and bonding pairs about the central atom
- the repulsion between those electron pairs
- why the molecule has the particular shape that it does and no other.

 ISBN: 9780170352611

3 Build a model of the NF_3 molecule and draw the 3D structure in the box below.

Explain whether the molecule is polar or not. In your answer you must include:

- the shape
- whether the shape contains polar bonds or not and how you know this
- whether the molecule is symmetrical or not
- why the molecule is polar or not.

4 Build a model of the IF_5 molecule and draw the 3D structure in the box below.

Explain whether the molecule is polar or not. In your answer you must include:

- the shape
- whether the shape contains polar bonds or not and how you know this
- whether the molecule is symmetrical or not
- why the molecule is polar or not.

Forces of attraction between particles

There are six types of attraction between particles that you will need to know about:

1. Ionic
2. Metallic
3. Covalent
4. Instantaneous dipole-induced dipole
5. Permanent dipole-dipole
6. Hydrogen.

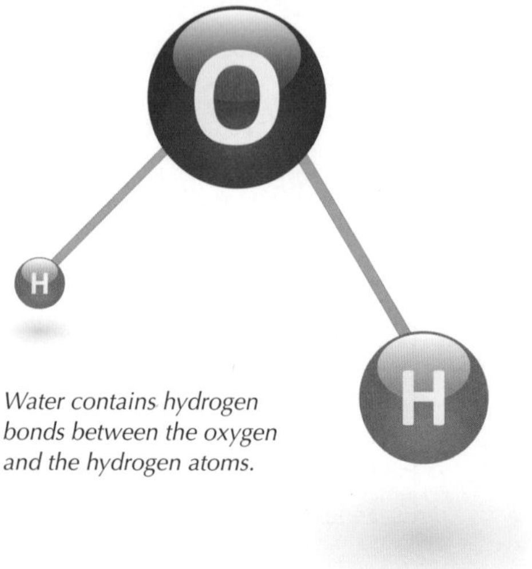

Water contains hydrogen bonds between the oxygen and the hydrogen atoms.

1 Ionic bonds

This electrostatic attraction occurs between two oppositely charged ions. It is strong and requires a large amount of energy to break apart. Ionic solids have a rigid lattice structure when solid, which does not allow the ions to move; however, once molten or aqueous (dissolved in water), the ions become free to move and they can conduct a current. The image to the right shows one of the most common ionic solids, sodium chloride, or table salt.

2 Metallic bonds

This electrostatic attraction occurs between the metal nuclei and the valence electrons which are delocalised and so free to move. As these electrons are free to move, metallic solids conduct both electricity and heat well, as well as being ductile and malleable. Metallic solids require a large amount of energy to break apart as the metallic bonds are so strong, as shown by the picture to the left as high temperatures are used to weld pieces of metal together.

 ISBN: 9780170352611

3 Covalent bonds

This attraction occurs between two non-metal atoms which can share one or more pair of electrons. It can be either polar (where the electron pair is shared unequally due to a difference in electronegativity between the atoms) or non-polar (where the electron pair in the bond is shared equally as there is very little or no difference between the electronegativity between the atoms). These bonds are extremely strong and require large amounts of energy to break them apart, however they do not have the potential of free-moving charged particles and so will not conduct a current (apart from graphite, as one electron is free to move on each carbon atom while the other three are connected by covalent bonds). Diamond, which is shown above, is made up of only carbon-to-carbon covalent bonds giving it its strength as the hardest known substance on Earth.

4 Instantaneous dipole-induced dipole bonds

This type of interaction occurs because moving electrons can, for an instant, be concentrated at one end of the molecule. Now it is a dipole and can induce a dipole on a neighbouring molecule. So, all particles with moving electrons have these forces. A dipole is a separation of charge caused by a bonded pair of electrons being closer to one atom in the bond. Larger molecules are more readily polarised. That is why the strength of these forces depends on particle size.

CHEMISTRY APPS

Geckos form instantaneous dipole-induced dipole bonds between the fine hair structures on their feet and a wall. These bonds are strong enough to hold the gecko to the wall but weak enough that the gecko can break them before moving.

5 Permanent dipole-dipole bonds

This type of interaction is a bond that occurs between polar molecules where the dipoles are always present, making this type of attraction stronger than instantaneous dipole-induced dipole ones.

EXAMINER'S TIP

These bonds, either instantaneous dipole-induced dipole or permanent dipole, will break when a substance containing them melts or boils, not the covalent bonds which are inside the molecule. However, only a few of them will break when a substance melts, but all of them will break when a substance boils.

6 Hydrogen bonds

This is a type of bond that can exist between molecules. A non-bonding pair of electrons on O (or N or F) is attracted to the hydrogen atom of a neighbouring molecule. This occurs between oxygen and hydrogen, nitrogen and hydrogen, or fluorine and hydrogen. They provide molecules with these bonds occurring within them with slightly higher than expected boiling points.

In water, the hydrogen bond provides it with the unique ability for its solid state to be less dense than its liquid state. Which is handy, as if it wasn't, ice would freeze over water sources from the bottom up causing lakes and seas to be permanently frozen over as they would never get the chance to fully melt or harbour life during the winter months.

1.10: Forces of attraction

a List the type of interactions in the following substances.

i HF ______________________

ii CO_2 ______________________

iii graphite ______________________

iv NaCl ______________________

v Mg ______________________

b Explain why H_2O has a much higher boiling point than H_2S, in terms of the nature of the bonds present in its structure.

____

____

____

____

____

____

____

____

____

____

____

ISBN: 9780170352611

c

Molecule	Boiling point (°C)
NH_3, ammonia	-33.3
PH_3, phosphine	-87.7

Explain the difference in the boiling points shown above in terms of the nature of the bonds present in its structure.

d

Molecule	Melting point (°C)
CH_4, methane	-182
C_5H_{12}, pentane	-130
$C_{10}H_{22}$, decane	-30.5

Explain the difference in the boiling points shown above in terms of the nature of the bonds present in its structure.

What do the following key words mean?

Ionic bond	
Metallic bond	
Covalent bond	
Instantaneous dipole-induced dipole bond	
Permanent dipole-dipole bond	
Polar bond	
Hydrogen bond	
Dipole	

ISBN: 9780170352611

EXPERIMENT 1

The effect of the hydrogen bond

AIM: To investigate the effects of hydrogen bonding on the change in enthalpy, ΔH, when mixing two liquids.

EQUIPMENT AND CHEMICALS:

ethanol, C_2H_5OH 10 mL and 25 mL measuring cylinder cyclohexane, C_6H_{12}
thermometer

SAFETY PRECAUTIONS:

Ethanol is highly flammable and an eye irritant. Cyclohexane is both a skin and eye irritant as well as being highly flammable.

METHOD:

1 Measure 10 mL of water into a 25 mL measuring cylinder and record its initial temperature.
2 Add 10 mL of ethanol to the water by measuring it in a separate measuring cylinder first and record the temperature change. And then, add it to the water.
3 Remove the thermometer and record the final volume.
4 Repeat the above steps, but use cyclohexane instead of water.

RESULTS:

Water and ethanol	
Temperature of water (°C)	
Temperature of mixture (°C)	
Total volume (mL)	

Ethanol and cyclohexane	
Temperature of cyclohexane (°C)	
Temperature of mixture (°C)	
Total volume (mL)	

QUESTIONS:

Explain why both the temperature and the volume were altered when water was mixed with ethanol but not when ethanol was mixed with cyclohexane in terms of the nature and structure of the substances.

ISBN: 9780170352611

Lewis structures, shapes and types of bonds

1 a Draw the Lewis diagrams and state the shapes for the molecules below.

PCl_5	SF_4	XeF_4
Lewis structure	Lewis structure	Lewis structure
Shape	Shape	Shape

b Explain why SF_4 has a different shape than XeF_4 despite both of them containing four bonded atoms about the central atom.

2 Explain the trends in boiling points shown in the graph using the nature and structure of the bonds present.

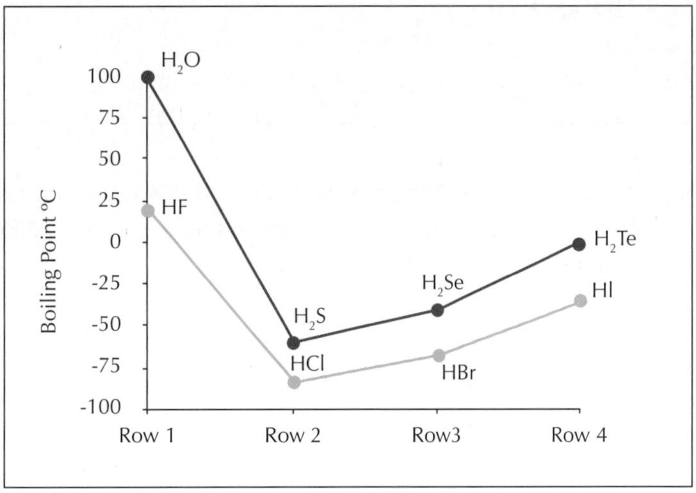

CHECKPOINT 2

WHAT HAVE YOU LEARNED SO FAR?

Specific heat capacity

This is the amount of energy required to raise the temperature of 1 gram of a substance by 1°C. Using this constant for a given substance, we can work out the amount of energy released or absorbed by a substance if the mass varies. To do this we use the following formula:

$$q = mc\Delta T$$

q = the change in energy (J)
m = mass (g)
c = the specific heat capacity ($Jg^{-1}{}^{o}C^{-1}$)
ΔT = the change in temperature (°C)

Most of the questions you will be asked to do with specific heat capacity will be in a solution, therefore the specific heat capacity will usually be given for water, as solutions are mostly water and the mass will be the mass of the solution.

Example: How much energy is released if 45 mL of sodium hydroxide solution is added to 90 mL of hydrochloric acid solution and the temperature of the mixture increases from 20°C to 24°C? (1 mL of water = 1 g)

The specific heat capacity of water = 4.18 $Jg^{-1}\,{}^{o}C^{-1}$

$q = (90 + 45) \times 4.18 \times (24 - 20)$
$= 135 \times 4.18 \times 4$
$= 2260$ J (3 s.f.)

It is worthwhile noting that the change in temperature (ΔT) must always be a positive number, since you cannot get negative energy. Exothermic reactions have a negative enthalpy because it has different units ($kJmol^{-1}$) and the negative sign represents the fact that the reaction has lost energy to the surroundings.

 ISBN: 9780170352611

1.11: Specific heat capacity

a How much heat is required to raise the temperature of 50 g of liquid water from 25°C to 90°C?

b What is the change in temperature of 30.2 g of liquid water if the heat required was 2500 J?

c How much heat is released in cooling 930 g of liquid water from 52°C to 18°C?

d i How much heat is released when 6.00 g of calcium oxide is completely dissolved in 250 mL of hydrochloric acid and the temperature of the solution rose from 20.0°C to 30.0°C?

ii If $M_r(CaO) = 56.0\ gmol^{-1}$, calculate Δ_rH for the reaction below using the formula $\Delta_rH = q/moles$.

$$CaO_{(s)} + 2HCl_{(aq)} \longrightarrow CaCl_{2\,(aq)} + H_2O_{(l)}$$

e A 38 g sample of water releases 1621 J of heat energy and cools from 34°C. What was the final temperature?

f i How much heat is released when 7.00 g of magnesium hydroxide is completely dissolved in 250 mL of nitric acid and the temperature of the solution rose from 21.0°C to 25.5°C?

ii If $M_r(Mg(OH)_2) = 40.3\ gmol^{-1}$, calculate Δ_rH for the reaction below using the formula $\Delta_rH = q/moles$.

$$Mg(OH)_{2\,(s)} + 2HNO_{3\,(aq)} \longrightarrow Mg(NO_3)_{2\,(aq)} + 2H_2O_{(l)}$$

FACT

Joseph Black, a Scottish chemist and physician, came up with idea for specific heat capacity in 1761 when he realised that different substances required different amounts of heat to change their temperature by a certain amount. This idea and others helped his friend James Watt create the steam engine.

Enthalpy, ΔH

Enthalpy is the measure of the heat content of a substance and is measured in $kJmol^{-1}$. In one mole of a reaction we can measure the change in enthalpy ($\Delta_r H$) that has taken place by measuring the change in temperature.

There are several different types of change in enthalpies you will need to be able to remember and write reactions for, as shown below.

Enthalpy of formation, $\Delta_f H^o$

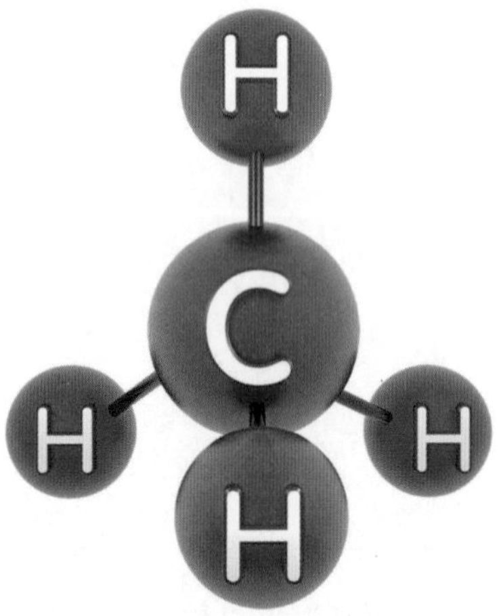

This is the amount of energy required to make ***one mole*** of a substance from its elements in their standard states.

Example:

$$C_{(s)} + 2H_{2\,(g)} \longrightarrow CH_{4\,(g)}$$

This is the formation of methane, CH_4, which has $\Delta_f H^o = -74.9\ kJmol^{-1}$

The states are vitally important to these equations because you will get a different value of $\Delta_f H^o$ depending on the state of the product, so make sure you include them. Also, you can form only *one mole of product,* so you may need to use fractions in order to balance these equations. The 'o' symbol indicates that the enthalpy was measured under standard conditions of room temperature and pressure.

Note: The enthalpy of formation for any element in its standard state is zero.

1.12: Enthalpy of formation

Write the equations for the enthalpy of formation for the following substances.

a H_2O (l) ____________________

b CO_2 (g) ____________________

c C_5H_{12} (l) ____________________

d MgO (s) ____________________

e $CaCO_3$ (s) ____________________

Enthalpy of combustion, $\Delta_c H^o$

The enthalpy of combustion is the amount of energy required to burn ***one mole*** of a substance in oxygen.

Example:

$$C_2H_5OH_{(l)} + 3O_{2\,(g)} \longrightarrow 2CO_{2\,(g)} + 3H_2O_{(g)}$$

Note: The H_2O is in a gaseous state due to the heat required to combust ethanol. Make sure you watch the states, as the enthalpy value will change depending on the state of any substance in the reaction.

Also, only one mole of ethanol is being burnt, so we need to have three moles of oxygen in order for the equation to be balanced.

ISBN: 9780170352611

1.13: Enthalpy of combustion

Write the equations for the complete combustion of the following substances.

a H_2 (g) ______

b CH_4 (g) ______

c C_5H_{12} (l) ______

d $C_{12}H_{26}$ (s) ______

e C_3H_7OH (l) ______

Enthalpy of the change of state

There are three you need to know about:

1 **The enthalpy of fusion**, $\Delta_{fus}H^o$, is the amount of energy required to break the weakest bonds for one mole of a substance to go from solid to liquid states. For example:

$$H_2O_{(s)} \longrightarrow H_2O_{(l)}$$

2 The **enthalpy of vaporisation**, $\Delta_{vap}H^o$, is the amount of energy required to break the bonds in one mole of a substance to go from liquid to gaseous states. For example:

$$H_2O_{(l)} \longrightarrow H_2O_{(g)}$$

3 The **enthalpy of sublimation**, $\Delta_{sub}H^o$, is the amount of energy required to break the bonds in one mole of a substance to go from solid to gaseous states. For example:

$$H_2O_{(s)} \longrightarrow H_2O_{(g)}$$

1.14: Enthalpies of change of state

Write the equations for the enthalpy of change of state for the following substances.

a $\Delta_{vap}H^o(H_2O)$ ______

b $\Delta_{sub}H^o(H_2O)$ ______

c $\Delta_{vap}H^o(CO)$ ______

d $\Delta_{fus}H^o(CO_2)$ ______

e $\Delta_{sub}H^o(C_4H_{10})$ ______

ISBN: 9780170352611

Entropy

In order to work out if a reaction is spontaneous or not we need a guide, and this guide is **entropy**.

Entropy is a measure of the disorder or the dispersal of energy in a system. It has the symbol S. The disorder refers to the arrangement and kinetic energies of particles in a system.

Generally, change happens in the direction in which the entropy will increase.

Entropy can be altered by four things:

1. The temperature a system is in; generally the higher the temperature of a system, the more kinetic energy the particles have and so the more entropy a system has.
2. The number of particles that are in a system; the more particles, the more arrangements that can be made and so the more entropy.
3. The complexity of the system; the more complex the particles, the more entropy a system can have.
4. If a system is in a gaseous state, the more volume the system has, the more entropy it will also have.

CHEMISTRY APPS

Gradually, the universe is increasing its entropy, from the Big Bang until its eventual death called the heat death of the universe, where the maximum possible entropy has been realised. At this point, there will be no place in the universe that contains a higher concentration of energy than another. Black holes are regions of space where entropy has already hit its maximum.

An artist's impression of a black hole in space.

1.15: Entropy

Predict whether the entropy of the following systems increases or decreases as a results of these changes.

a $KCl_{(s)} + aq \longrightarrow KCl_{(aq)}$ ______

b $CO_{2\,(s)} \longrightarrow CO_{2\,(g)}$ ______

c $Ca_{(s)} + Cl_{2\,(g)} \longrightarrow CaCl_{2\,(s)}$ ______

d **i** $N_2O_{4\,(g)} \longrightarrow 2NO_{2\,(g)}$ ______

ii Explain why you think the entropy either increased or decreased for **i**.

______

______

e **i** $Na_2CO_{3\,(s)} + 2HCl_{(aq)} \longrightarrow 2NaCl_{(aq)} + H_2O_{(l)} + CO_{2\,(g)}$ ______

ii Explain why you think the entropy either increased or decreased for **i**.

______

______

ISBN: 9780170352611

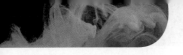

Hess's Law

Hess's Law allows you to calculate the enthalpy of reaction using an alternative pathway, as no matter how you get there, the enthalpy of reaction under standard conditions should be the same.

Germain Henri Hess
1802–50

Example: Calculate the $\Delta_r H$ for $Al_2O_{3\,(s)} + 3Mg_{\,(s)} \longrightarrow 2Al_{\,(s)} + 3MgO_{\,(s)}$ given the following information:

Equation one: $2Al_{\,(s)} + 1\tfrac{1}{2}O_{2\,(g)} \longrightarrow Al_2O_{3\,(s)}$ $\qquad \Delta_f H^o = -1676\ kJmol^{-1}$

Equation two: $Mg_{\,(s)} + \tfrac{1}{2}O_{2\,(g)} \longrightarrow MgO_{\,(s)}$ $\qquad \Delta_f H^o = -602\ kJ\ mol^{-1}$

In order to get these two equations to equal the unknown $\Delta_r H$, we must ensure we have the same number of moles of each reactant and product and we must make sure that reactants and products are on the same side.

So to equation one: $2Al_{\,(s)} + 1\tfrac{1}{2}\ O_{2\,(g)} \longrightarrow Al_2O_{3\,(s)}$ $\qquad \Delta_f H^o = -1676\ kJmol^{-1}$

The first thing we must do is reverse it so the Al is on the product side and the $Al_2O_{3\,(s)}$ is on the reactant side. When we do this to the equation, we must also do it to the $\Delta_f H^o$ by reversing the sign.

$$2Al_{\,(s)} + 1\tfrac{1}{2}O_{2\,(g))} \longleftarrow Al_2O_{3\,(s]} \qquad \Delta_f H^o = +1676\ kJmol^{-1}$$

To equation two: $Mg_{\,(s)} + \tfrac{1}{2}\ O_{2\,(g)} \longrightarrow MgO_{\,(s)}$ $\qquad \Delta_f H^o = -602\ kJmol^{-1}$

We will need to multiply it by 3 so that there are the same number of moles of both Mg and MgO, which means the $\Delta_f H^o$ must also be multiplied by 3.

$$3Mg_{\,(s)} + 1\tfrac{1}{2}O_{2\,(g)} \longrightarrow 3MgO_{\,(s)} \qquad \Delta_f H^o = -602\ kJmol^{-1} \times 3 = -1806\ kJmol^{-1}$$

Finally, add up all the $\Delta_f H^o$ for both equation one and two in order to find the $\Delta_r H$.

$$\Delta_r H = +1676 - 1806 = -130\ kJmol^{-1}$$

There is another way of completing this calculation using the formula shown below, but only if the equations given to you are *enthalpies of formation*.

$$\Delta_r H = \Sigma\Delta_f H^o\ (products) - \Sigma\Delta_f H^o\ (reactants)$$

Example: So for the above example you would do the following:

$\Delta_r H = \Sigma\Delta_f H^o\ (MgO) \times 3 - \Sigma\Delta_f H^o\ (Al_2O_3)$

$= -602 \times 3 - {}^{-}1676$

$= -130\ kJmol^{-1}$

As the $\Delta_f H^o$ for any element is zero, you can leave out the other reactant and product.

We still must multiply equation two by 3 in order to get the right number of moles.

1.16: Hess's Law

a Calculate ΔH for the reaction $C_2H_{4\,(g)} + H_{2\,(g)} \longrightarrow C_2H_{6\,(g)}$ using the following data.

$C_2H_{4\,(g)} + 3O_{2\,(g)} \longrightarrow 2CO_{2\,(g)} + 2H_2O_{(l)}$ $\Delta H = -1410\ kJmol^{-1}$

$C_2H_{6\,(g)} + 3\frac{1}{2}O_{2\,(g)} \longrightarrow 2CO_{2\,(g)} + 3H_2O_{(l)}$ $\Delta H = -1560\ kJmol^{-1}$

$H_{2\,(g)} + \frac{1}{2}O_{2\,(g)} \longrightarrow H_2O_{(l)}$ $\Delta H = -286\ kJmol^{-1}$

b Calculate ΔH for the reaction $4NH_{3\,(g)} + 5O_{2\,(g)} \longrightarrow 4NO_{(g)} + 6H_2O_{(g)}$ using the following data.

$N_{2\,(g)} + O_{2\,(g)} \longrightarrow 2NO_{(g)}$ $\Delta H = -181\ kJmol^{-1}$

$N_{2\,(g)} + 3H_{2\,(g)} \longrightarrow 2NH_{3\,(g)}$ $\Delta H = -91.8\ kJmol^{-1}$

$2H_{2\,(g)} + O_{2\,(g)} \longrightarrow 2H_2O_{(g)}$ $\Delta H = -484\ kJmol^{-1}$

c Find $\Delta_f H^\circ$ for ethanoic acid, CH_3COOH, using the following thermochemical data.

$CH_3COOH_{(l)} + 2O_{2\,(g)} \longrightarrow 2CO_{2\,(g)} + 2H_2O_{(l)}$ $\Delta H = -875\ kJmol^{-1}$

$\Delta_f H(CO_2) = -395\ kJmol^{-1}$

$\Delta_f H(H_2O) = -286\ kJmol^{-1}$

(Hint: You will have to write the formation equations for CO_2 and H_2O before you can calculate it.)

ISBN: 9780170352611

d Calculate ΔH for the reaction $CH_{4\,(g)} + NH_{3\,(g)} \longrightarrow HCN_{(g)} + 3H_{2\,(g)}$ using the following reactions.

$N_{2\,(g)} + 3H_{2\,(g)} \longrightarrow 2NH_{3\,(g)}$ $\Delta H = -91.8\ kJmol^{-1}$

$\Delta_f H^o(CH_{4\,(g)}) = -74.9\ kJmol^{-1}$

$H_{2\,(g)} + 2C_{(s)} + N_{2\,(g)} \longrightarrow 2HCN_{(g)}$ $\Delta H = +270.3\ kJmol^{-1}$

e Calculate ΔH for the reaction $2Al_{(s)} + 3Cl_{2\,(g)} \longrightarrow 2AlCl_{3\,(s)}$ using the following data.

$2Al_{(s)} + 6HCl_{(aq)} \longrightarrow 2AlCl_{3\,(aq)} + 3H_{2\,(g)}$ $\Delta H = -1050\ kJmol^{-1}$

$HCl_{(g)} \longrightarrow HCl_{(aq)}$ $\Delta H = -74.8\ kJmol^{-1}$

$H_{2\,(g)} + Cl_{2\,(g)} \longrightarrow 2HCl_{(g)}$ $\Delta H = -1850\ kJmol^{-1}$

$AlCl_{3\,(s)} \longrightarrow AlCl_{3\,(aq)}$ $\Delta H = -323\ kJmol^{-1}$

What do the following key words mean?

Specific heat capacity	
Enthalpy	
Enthalpy of formation	
Enthalpy of combustion	
Enthalpy of fusion	
Enthalpy of sublimation	
Enthalpy of vaporisation	
Entropy	

ISBN: 9780170352611

EXPERIMENT 2

Hess's Law

AIM: To calculate the enthalpy change that occurs for the water of crystallisation for magnesium sulfate.

EQUIPMENT AND CHEMICALS:

magnesium sulfate anhydrous
thermometer
100 mL measuring cylinder
magnesium sulfate hydrated
electronic scales
polystyrene cup
stirring rod

SAFETY PRECAUTIONS:

Magnesium sulfate anhydrous and magnesium sulfate heptahydrate are mild irritants to skin, eyes and lungs.

METHOD:

1 Using the measuring cylinder, measure 100 mL of water and pour this water into the polystyrene cup. Then measure the temperature of the water.
2 Accurately weigh about **7.5 g** of anhydrous magnesium sulfate using the electronic balance. Add this to the water and stir using a stirring rod. Record the maximum change in temperature.
3 Rinse out the polystyrene cup and stirring rod and then pour a fresh 100 mL of water in the polystyrene cup and record its temperature.
4 Accurately weigh about **15.5 g** of magnesium sulfate heptahydrate using the electronic balance. Add this to the water and stir using a stirring rod. Record the maximum change in temperature.

RESULTS:

Temperature of water before adding the $MgSO_4$ (°C)	
Maximum change in temperature with the $MgSO_4$ (°C)	

Temperature of water before adding the $MgSO_4.7H_2O$ (°C)	
Maximum change in temperature with the $MgSO_4.7H_2O$ (°C)	

$M(MgSO_4.7H_2O)$ = ______________________ $gmol^{-1}$

$M(MgSO_4)$ = ______________________ $gmol^{-1}$

QUESTIONS:

1 Calculate ΔH for the following equation from the experimental values.

$$MgSO_{4\,(s)} + 7H_2O_{(l)} \longrightarrow MgSO_4.7H_2O_{(s)}$$

(Hint: You will need to work out the ΔH of each of the above two experiments using the formulas $q = mc\Delta T$ and $\Delta H = q/moles$ and then use Hess's Law.)

 ISBN: 9780170352611

2 Calculate ΔH for the above equation using the theory values given to you below.

$MgSO_{4\,(s)} \longrightarrow Mg^{2+}_{(aq)} + SO^{2-}_{4\,(aq)}$ ΔH = -84 $kJmol^{-1}$

$MgSO_4.7H_2O_{(s)} \longrightarrow Mg^{2+}_{(aq)} + SO^{2-}_{4\,(aq)}$ ΔH = +16 $kJmol^{-1}$

3 Compare and contrast the values you calculated for ΔH in both questions **1** and **2**. Why might they be similar or different?

CHECKPOINT 3
WHAT HAVE YOU LEARNED SO FAR?

Specific heat capacity and Hess's Law

1 55 g of calcium hydroxide solution is added to 45 g of hydrochloric acid solution. The temperature of the mixture increases from 21.0°C to 25.5°C. Calculate the heat released from this experiment. (Remember, c = 4.18 $J^oC^{-1}g^{-1}$.)

$$Ca(OH)_{2\,(aq)} + 2HCl_{(aq)} \longrightarrow CaCl_{2\,(aq)} + H_2O_{(l)}$$

2 Calculate the energy needed to melt 75.0 g of ice at its melting point of 0°C using the following two pieces of information.

$M_r(H_2O) = 18\ gmol^{-1}$ $\Delta_{fus}H^o(H_2O) = 6.02\ kJmol^{-1}$

ISBN: 9780170352611

3 Calculate the ΔH value for the following reaction using the information given.

$$H_{2\,(g)} + S_{(s)} + 2O_{2\,(g)} \longrightarrow H_2SO_{4\,(g)}$$

$S_{(s)} + O_{2\,(g)} \longrightarrow SO_{2\,(g)}$ $\Delta H = -297\ kJmol^{-1}$

$SO_{2\,(g)} + H_2O_{(g)} + \frac{1}{2}O_{2\,(g)} \longrightarrow H_2SO_{4\,(g)}$ $\Delta H = -900\ kJmol^{-1}$

$H_{2\,(g)} + \frac{1}{2}O_{2\,(g)} \longrightarrow H_2O_{(g)}$ $\Delta H = -242\ kJmol^{-1}$

4 Calculate the ΔH value for the following reaction using the information given.

$$CH_{4\,(g)} + 2O_{2\,(g)} \longrightarrow CO_{2\,(g)} + 2H_2O_{(g)}$$

$C_{(s)} + 2\,H_{2\,(g)} \longrightarrow CH_{4\,(g)}$ $\Delta H = -75\ kJmol^{-1}$

$CO_{2\,(g)} \longrightarrow C_{(s)} + O_{2\,(g)}$ $\Delta H = 393\ kJmol^{-1}$

$H_{2\,(g)} + \frac{1}{2}O_{2\,(g)} \longrightarrow H_2O_{(g)}$ $\Delta H = -242\ kJmol^{-1}$

ISBN: 9780170352611

EXAM-TYPE QUESTIONS

Before you get started here are some points to remember from the Examiner's Report:

- Don't use generalisations; use terms carefully.
- Plan your answer before you start.

Question one

a Write the electron configuration using *s, p, d, f* notation for:

Cu ______________________

Fe^{3+} ______________________

b Describe and explain the trend in atomic radius down group 17 — F, Cl, Br, I — of the periodic table.

c Explain why the first ionisation energy for bromine is less than that of chlorine.

d Explain why the second ionisation energy of sodium is much higher than the first.

First ionisation energy of sodium ($kJmol^{-1}$)	496
Second ionisation energy of sodium ($kJmol^{-1}$)	4563

ISBN: 9780170352611

Question two

a Write the electron configuration using *s, p, d, f* notation for:

Cl

Cr^{3+}

b Explain the trend shown in the graph for atomic radii in period 3.

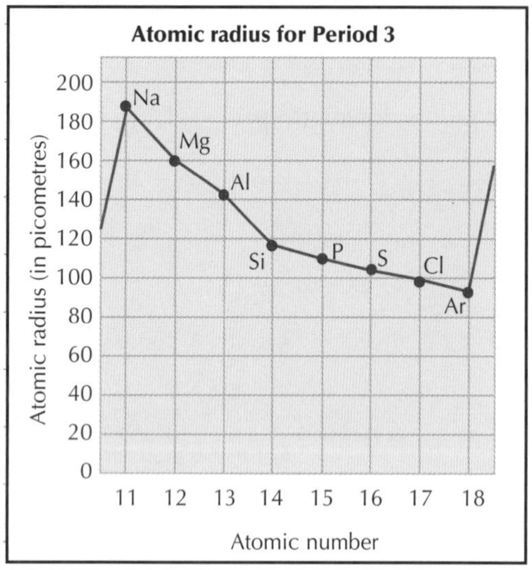

c Explain the trend in electronegativity as you go across period 3.

d Explain the difference between the size of a fluorine atom and the size of a fluoride ion.

Fluorine atom size (picometres)	71
Fluoride ion size (picometres)	119

 ISBN: 9780170352611

EXAM-TYPE QUESTIONS

Question three

a Explain why aluminium has a greater atomic radius than phosphorus but a smaller ionic radius.

Atom	Atomic radius (pm)	Ion	Ionic radius (pm)
Al	118	Al^{3+}	68
P	110	P^{3-}	180

b Explain the difference in first ionisation energies of B, C and N.
First ionisation energies:

B	807 $kJmol^{-1}$
C	1093 $kJmol^{-1}$
N	1402 $kJmol^{-1}$

Question four

a Complete the table below drawing Lewis diagrams and determining the shape for each ion.

	SF_3^-	SO_4^{2-}	SO_3^{2-}
Lewis diagrams			
Shape			

b Discuss the reasons for the difference in shape of the ions SF_3^-, SO_4^{2-} and SO_3^{2-}.

 ISBN: 9780170352611

EXAM-TYPE QUESTIONS

Question five

a Complete the table below drawing Lewis diagrams, determining the shape and polarity for each molecule.

	PCl_5	ICl_5
Lewis diagrams		
Shape		
Polar or non-polar		

b Discuss the reasons for the difference in shape and polarity of PCl_5 and ICl_5.

ISBN: 9780170352611

Question six

a Complete the table below drawing Lewis diagrams, determining the shape, shape diagram and polarity for each molecule.

	SF_4	XeF_4
Lewis diagrams		
Shape and shape diagram		
Polar or non-polar		

b Discuss the reasons for the difference in shape and polarity of SF_4 and XeF_4.

 ISBN: 9780170352611

EXAM-TYPE QUESTIONS

Question seven

a Define $\Delta_{sub}H^o$.

b Standard enthalpy of reaction ($\Delta_r H^o$) refers to the reaction performed under standard conditions. What are the standard conditions?

c Given the thermodynamic data below, calculate the ΔH value for the reaction:

$$2C_{(s)} + 3H_{2\,(g)} + \tfrac{1}{2}O_{2\,(g)} \longrightarrow C_2H_5OH_{(g)}$$

$C_2H_5OH_{(g)} + 3O_{2\,(g)} \longrightarrow 2CO_{2\,(g)} + 3H_2O_{(g)}$ ΔH = -1368 kJmol^{-1}

$CO_{2\,(g)} \longrightarrow C_{(s)} + O_{2\,(g)}$ ΔH = 393 kJmol^{-1}

$H_{2\,(g)} + \tfrac{1}{2}O_{2\,(g)} \longrightarrow H_2O_{(g)}$ ΔH = -242 kJmol^{-1}

Question eight

a Explain what $\Delta_f H^o$ means.

b Write the $\Delta_f H^o$ equation for both $CO_{2\ (g)}$ and $H_2O_{(g)}$.

c Butane burns according to the equation:

$$C_4H_{10\ (l)} + 1\frac{1}{2}O_{2\ (g)} \longrightarrow 4CO_{2\ (g)} + 5H_2O_{(g)}$$

Use the following information to calculate $\Delta_r H^o$ for the above reaction.

$4C_{(s)} + 5H_{2\ (g)} \longrightarrow C_4H_{10\ (l)}$ $\Delta H^o = -126\ kJmol^{-1}$

$\Delta_f H^o(CO_{2\ (g)}) = -394\ kJmol^{-1}$

$\Delta_f H^o(H_2O_{(g)}) = -242\ kJmol^{-1}$

ISBN: 9780170352611

EXAM-TYPE QUESTIONS

Question nine

a i Calculate the final temperature when 0.55 g of hydrochloric acid is mixed with 0.55 g of sodium hydroxide solution, which had an initial temperature of 21.0°C and it released 2760 J of energy.

ii If $M_r(HCl) = 36.5\ gmol^{-1}$, calculate the $\Delta_r H$ for the reaction below using the formula $\Delta_r H = q/moles$.

$$NaOH_{(aq)} + HCl_{(aq)} \longrightarrow NaCl_{(aq)} + H_2O_{(l)}$$

b Calculate the enthalpy of the following reaction using the information provided:

$$3HCl_{(aq)} + HNO_{3\ (aq)} \longrightarrow Cl_{2\ (g)} + NOCl_{(g)} + 2H_2O_{(l)}$$

$\Delta_f H^\circ$ HCl = -92 $kJmol^{-1}$

$\Delta_f H^\circ$ HNO_3 = -174 $kJmol^{-1}$

$\Delta_f H^\circ$ NOCl = +52 $kJmol^{-1}$

$\Delta_f H^\circ$ H_2O = -286 $kJmol^{-1}$

Question ten

a The boiling points and enthalpies of vaporisation of ammonia and methane are shown in the table below.

Substance		Boiling point (°C)	$\Delta_{vap}H^{o}$ (kJmol⁻¹)
ammonia	NH_3	-33	23.4
methane	CH_4	-162	8

i Write the equation for which the enthalpy change is the enthalpy of vaporisation, $\Delta_{vap}H^{o}$, for methane.

ii Identify the types of intermolecular forces present in ammonia and methane and explain why ammonia has a higher boiling point than methane.

iii Discuss, in terms of intermolecular forces, why ammonia has a higher enthalpy of vaporisation than methane.

iv The enthalpy value for the vaporisation of methane is positive. Explain why the enthalpy of the reaction is positive and what will happen to the entropy of the system as the reaction proceeds.

 ISBN: 9780170352611

3.5 Demonstrate understanding of the properties of organic compounds (91391)

Learning outcomes

Tick off when you have studied these ideas in class and when you have revised that section prior to your assessment.

	Learning outcomes	In class	Revision
1	Describe and explain geometric, structural and optical isomers.		
2	Name, describe the properties of and reactions of the hydrocarbons — alkanes and alkenes.		
3	Name, describe the properties of and reactions of the alcohols.		
4	Name, describe the properties of and reactions of the haloalkanes.		
5	Name, describe the properties of and reactions of the amines.		
6	Name, describe the properties of and reactions of the aldehydes and ketones.		
7	Name, describe the properties of and reactions of the carboxylic acids and acid chlorides.		
8	Name, describe the properties of and reactions of the esters and amides.		
9	Draw the monomers and polymers of addition and condensation polymers including proteins.		

ISBN: 9780170352611

Pre-test — What do you know?

1 Fill in the table below.

	Name	Condensed structural formula	Molecular and empirical formula	Functional group
a	2,2-dimethylbutane			Alkane
b		$CH_3CH(Cl)CH_3$	C_3H_7Cl	
c		CH_3COOH		
d	2-methylpropan-1-ol			

2 Complete the reaction scheme shown below.

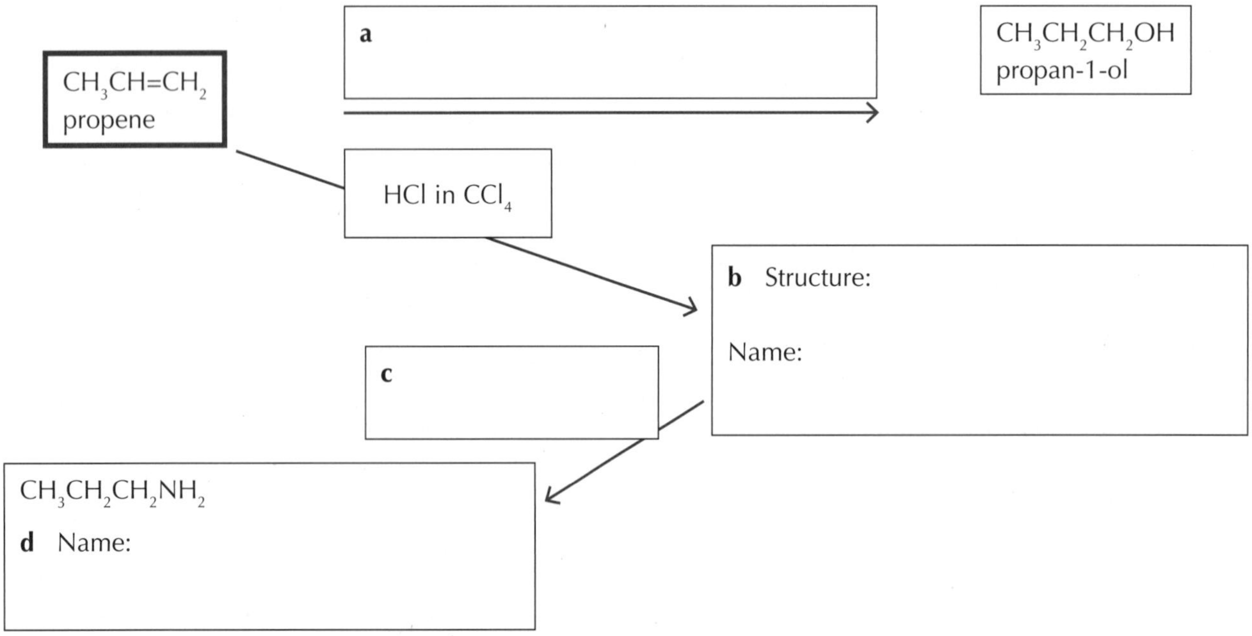

3 Draw and name all the structural isomers of C_5H_{12}.

ISBN: 9780170352611

Recapping Level 2

In Level 2 you learned how to name, the properties, the structures and the reactions of the following groups shown in the table below.

(Note that below is just an example for each family; there are more detailed notes throughout the chapter, as you must know all the reactions of these groups from last year as well as a few new ones.)

Family	Name	Structure	Properties	Reactions
Alkane	Methane	CH_4	Highly combustible	Undergo substitution reactions with halogens under UV light: $CH_3CH_3 + Cl_2 \xrightarrow{UV} CH_3CH_2Cl$
Alkene	Ethene	$CH_2{=}CH_2$	Insoluble in polar solvents	Undergo addition reactions with halogens: $CH_2CH_2 + Cl_2 \rightarrow CH_2ClCH_2Cl$
Alcohol	Ethanol	CH_3CH_2OH	Soluble up to 4 carbons in polar solvents	Undergo substitution reactions to form haloalkanes or amines: $CH_3CH_2OH + PCl_3 \rightarrow CH_3CH_2Cl$
Haloalkane	2-chloropropane	$CH_3CH(Cl)CH_3$	Not very soluble in polar solvents	Undergo substitution reactions to form amines or alcohols: $CH_3CH_2Cl + KOH_{(aq)} \rightarrow CH_3CH_2OH$
Amine	Propan-1-amine	$CH_3CH_2CH_2NH_2$	Basic	Undergo neutralisation reactions with acids to produce neutral salts: $CH_3CH_2NH_2 + HCl \rightarrow CH_3CH_2NH_3^+Cl^-$
Carboxylic acid	Propanoic acid	CH_3CH_2COOH	Acidic	Undergo neutralisation reactions with bases to form neutral salts: $CH_3COOH + NaOH \rightarrow CH_3COO^-Na^+ + H_2O$

Isomerism

Structural isomerism is when organic molecules have the same molecular formula but the atoms are arranged differently in space. There are two types you need to know:

1 Geometric isomerism

This occurs when you have two different groups attached to each carbon in a double bond, and because there is no free rotation about a double bond, two isomers form. One is called the 'cis' isomer, meaning that two of the same or similar groups are attached to one side of each carbon in the double bond, and the other is the 'trans' isomer, where the same or similar groups are attached to opposite sides. For example:

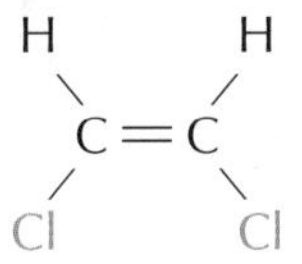

cis-1,2-dichloroethene

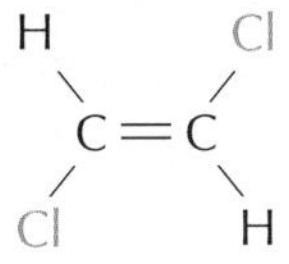

trans-1,2-dichloroethene

2 Optical isomerism

Optical isomerism involves an asymmetric carbon atom — a carbon atom bonded to four different atoms or groups of atoms. A molecule with an asymmetric carbon atom is known as a chiral molecule. The two forms of the chiral molecule are known as ***enantiomers***, or ***optical isomers***. They are mirror images when in crystal form of each other and cannot be superimposed onto each other. They can be separated out by the fact that they will each rotate a plane of polarised light (that is, light travelling in one direction only) in different directions.

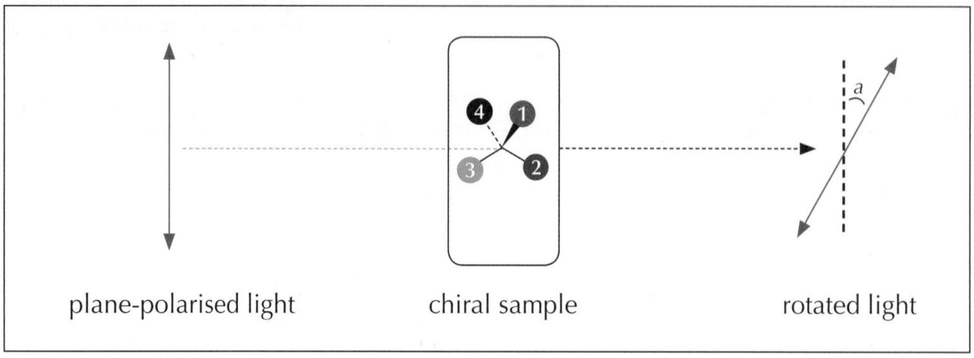

Example:

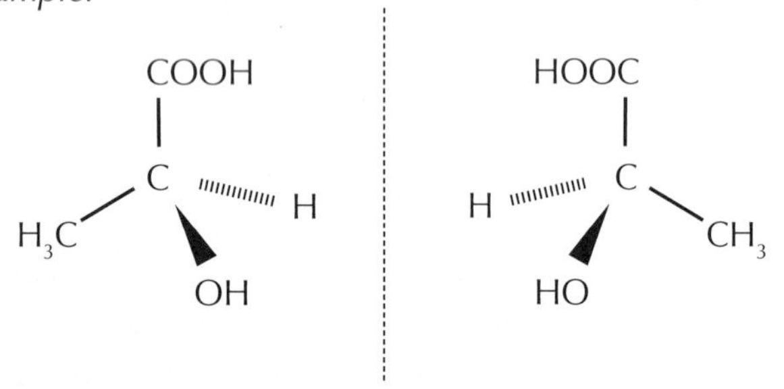

EXAMINER'S TIP

You need to know how to distinguish between two enantiomers both physically and experimentally for the assessment and you should be able to draw a diagram like this.

FACT

The compound carvone is largely responsible for both the taste and the smell of spearmint and caraway seeds — the difference between them is that they are optical isomers.

2.1: Isomers

1 Explain why but-1-ene cannot form geometric isomers but is a structural isomer of but-2-ene, which can form them. In your answer you should include:

- a definition for a structural isomer
- a definition for a geometric isomer
- structural formulas of each isomer
- an explanation of why one is an isomer of the other
- an explanation of why but-2-ene can form a geometric isomer.

ISBN: 9780170352611

2 **a** Which of the two molecules below contains a chiral carbon? (Put a * by the chiral carbon — carbon with four different groups attached.)

i

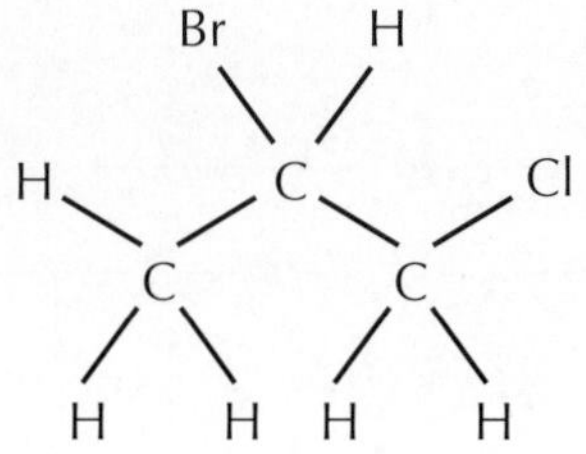

ii

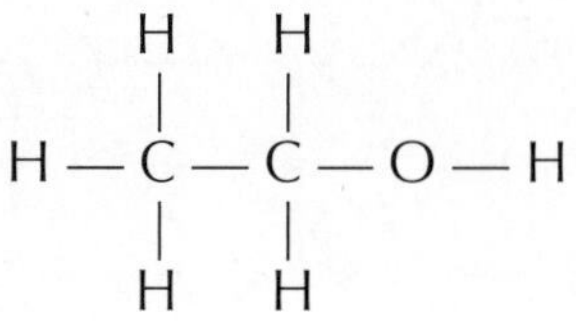

b Draw the 3D drawings of the two different enantiomers.

c Explain why the one you chose to have a chiral carbon can exist as optical isomers and how you could distinguish between the two isomers.

3 **a** Which of the two molecules below contains a chiral carbon? (Put a * by the chiral carbon.)

i

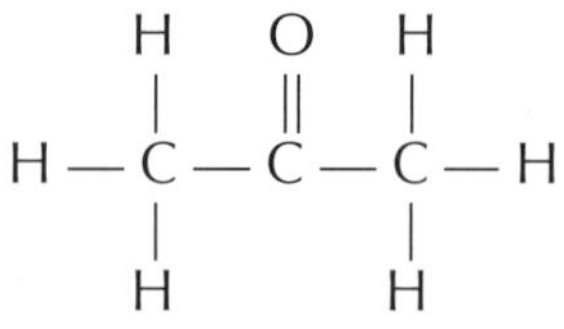

ii

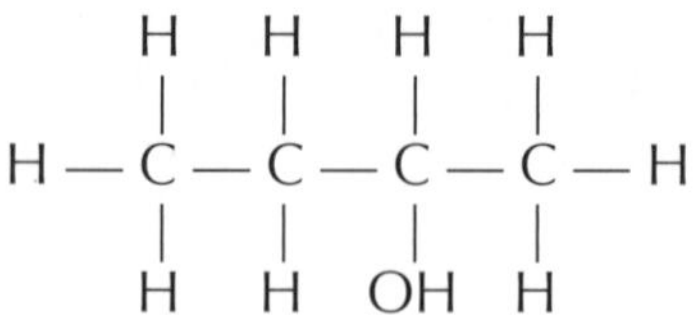

b Draw the 3D drawings of the two different enantiomers.

c Explain why you chose the molecule that you did to have a chiral carbon and how you could distinguish between the two isomers.

FACT

Louis Pasteur was the first person to isolate two enantiomers: these were the enantiomers of tartaric acid. He painstakingly sat there with a pair of tweezers and a hand lens and separated out the two isomers noticing that they were mirror images of each other. He then dissolved each in water and saw that one rotated a plane of polarised light to the right and the other to the left.

A butane stove top.

Hydrocarbons

Naming

Alkanes are named using two parts: a prefix that tells you how many carbons it contains, along with the suffix 'ane' on the end. Any branches coming off the main chain are numbered to give the lowest possible number.

Examples:

CH_4 is called methane (meth is for one carbon).

CH_3CH_3 is called ethane (eth is for two carbons, then comes prop, but, pent, hex, hept and oct).

$CH_3CH(CH_3)CH_2CH_3$ is called 2-methylbutane.

 ISBN: 9780170352611

Alkenes are also named using the same prefixes (meth, eth, …) followed by a number of what carbon the double bond is off (if the chain has more than four carbons present in it) and the suffix 'ene'.
Examples:
CH_2CH_2 is called ethene.
$CH_2CHCH_2CH_3$ is called but-1-ene.
$CH_2CHCH(CH_3)CH_3$ is called 3-methylbut-1-ene (the double bond gets the priority with numbering since it is the reactive part of the molecule).

The molecules above have been drawn using condensed structural formula; full structural formula shows all the bonds drawn out, for example ethene:

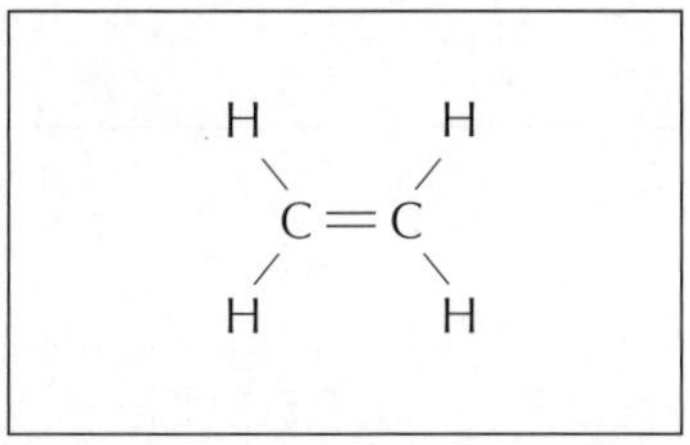

EXAMINER'S TIP

Make sure when you draw out organic structures, you draw in all hydrogens and put no more than four bonds around each carbon atom.

Physical properties

Alkanes and alkenes have similar physical properties:

- They are non-polar covalent molecules.
- They contain instantaneous dipole-induced dipole bonds between the molecules.
- They form waxy or greasy solids due to these weak intermolecular forces, as they require little energy to break.
- They have low melting and boiling points.
- They are insoluble in polar solvents like water, but will dissolve in non-polar solvents like tetrachloromethane. This is due to the attraction between the solvent (in a non-polar solvent) and the hydrocarbon being strong enough to overcome the intermolecular bonds in the hydrocarbon and the solvent.

Chemical properties

Alkanes undergo one type of reaction that is called **substitution**, where one hydrogen is swapped for a halogen (F, Cl, Br or I). It requires UV light as a catalyst and is a very slow reaction. For example:

$$CH_3CH_2CH_3 + Br_2 \xrightarrow{UV} CH_3CH_2CH_2Br$$

Alkenes undergo **addition** reactions where the double bond is lost and a small molecule is gained on the carbons in the double bond.

These reactions are given below, along with what is added on and any conditions required:

- hydrogenation — H_2 (need Ni and 300°C)
- bromination — Br_2 (with the colour change orange/brown to colourless)
- chlorination — Cl_2
- hydration — H-OH
- hydrohalogenation — HCl in CCl_4, HBr.

ISBN: 9780170352611

Remember Markovnikov's rule — *'The rich get richer.'* This means that a carbon that was originally richer in hydrogen will gain another hydrogen more commonly forming the major product when adding HCl, HBr or H_2O. For example:

$$CH_2CHCH_2CH_3 + HBr \longrightarrow \underset{\text{major product}}{CH_3CH(Br)CH_2CH_3} + \underset{\text{minor product}}{CH_2(Br)CH_2CH_2CH_3}$$

Alkenes also undergo **addition polymerisation** reactions where one alkene is added on to itself many times over. These reactions use heat, pressure and a catalyst which depends on the type of polymer that is being made. For example:

$$n\ CH_3CHCH_2 \longrightarrow \left[\begin{array}{c} CH_3 \\ | \\ CH - CH_2 \end{array}\right]_n$$

The polymer formed above is called polypropylene and is used in thermal clothing.

Finally, alkenes undergo **oxidation** using acidified potassium permanganate, $H^+/KMnO_4$, to form a diol. For example:

$$CH_2CH_2 + H^+/KMnO_4 \longrightarrow \underset{\text{ethan-1,2-diol}}{CH_2(OH)CH_2OH}$$

2.2: Hydrocarbons

1 Name or draw the condensed structural formula for the following molecules:

a $CH_3CH_2CH_3$ ____________________

b $CH_3C(CH_3)_2CH_3$ ____________________

c $CH_3CH(CH_3)CH_2CH_3$ ____________________

d 2,4-dimethylpentane ____________________

e 2-methylhexane ____________________

f $CH_3CH{=}CH_2$ ____________________

g $CH_3CH{=}C(CH_3)_2$ ____________________

h $CH_3CH{=}C(CH_3)CH_2CH_3$ ____________________

2 Write the reactions for the following:

a bromination of ethene

____

b hydration of propene

____

ISBN: 9780170352611

c the substitution of ethane with bromine

____

d the polymerisation of but-1-ene

____

e oxidation of but-2-ene (equation doesn't have to be balanced)

____

CHEMISTRY APPS

It is safer to have a full tank of petrol in your car than a half tank, as the volatile liquid will ignite much more readily in the gas state (which hovers above the liquid state), which readily forms once the gas tank starts to empty out.

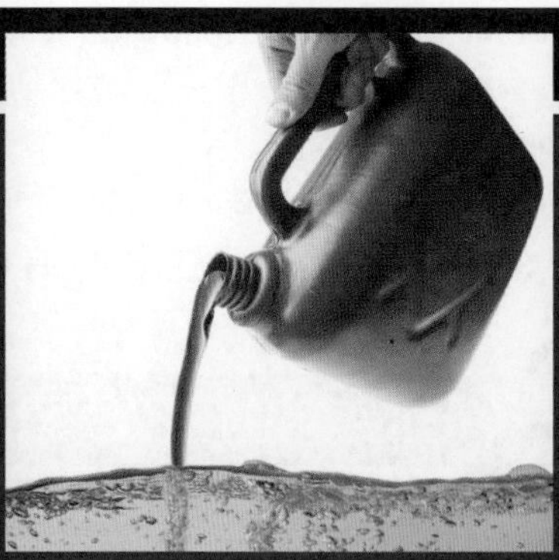

Alcohols

Naming

Alcohols are also named using the same prefix as hydrocarbons — meth, eth, prop, ..., then numbering what carbon the functional group is on followed by the suffix 'ol'. The alcohol group gets the priority in naming over any other branches, since it is the reactive part of the molecule (the functional group). The alcohol in the picture is ethanol; this is the only type of alcohol humans can drink.

Example:

$CH_3CH(CH_3)CH_2CH_2OH$ is called 3-methylbutan-1-ol.

Physical properties

Alcohols have higher melting and boiling points than their corresponding hydrocarbon due to the presence of the hydrogen bond at the OH end of the molecule, which forms strong bonds between molecules that require large amounts of energy to break.

They are readily water soluble with up to four carbons in the chain and slightly soluble with five or six carbons in the chain, after which they become insoluble. The water-soluble alcohols form strong hydrogen bonds with water, which overcomes the forces of attraction between the alcohol molecules. The insoluble alcohols have more instantaneous dipole-induced dipole forces due to the length of the carbon chain.

We can classify alcohols into three categories depending on how many carbons the carbon connected to the functional group has:

ISBN: 9780170352611

Primary	Secondary	Tertiary
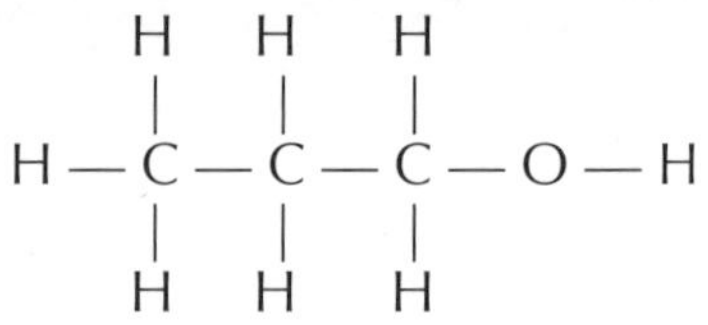	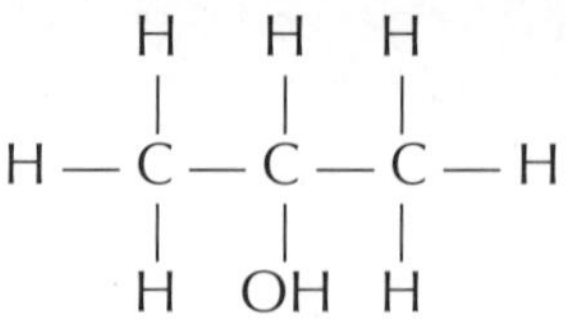	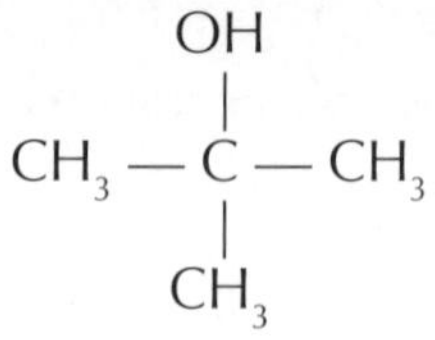

Chemical properties

Alcohols undergo the following reactions:

Substitution

Chlorination — $SOCl_2$ and reflux forming SO_2 as the other product. For example:

$CH_3CH_2OH + SOCl_2 \longrightarrow CH_3CH_2Cl + SO_2$

EXAMINER'S TIP

Reflux is used for two things in an organic reaction: to increase the rate of reaction, and to prevent the loss of volatile organic reactants and/or products. It is basically just boiling something gently.

The equipment for reflux is shown below. The condenser is where the boiled organic substance cools and flows back down into the reaction vessel ensuring no volatile organic is lost during the process; it does this by having cold water running through it constantly.

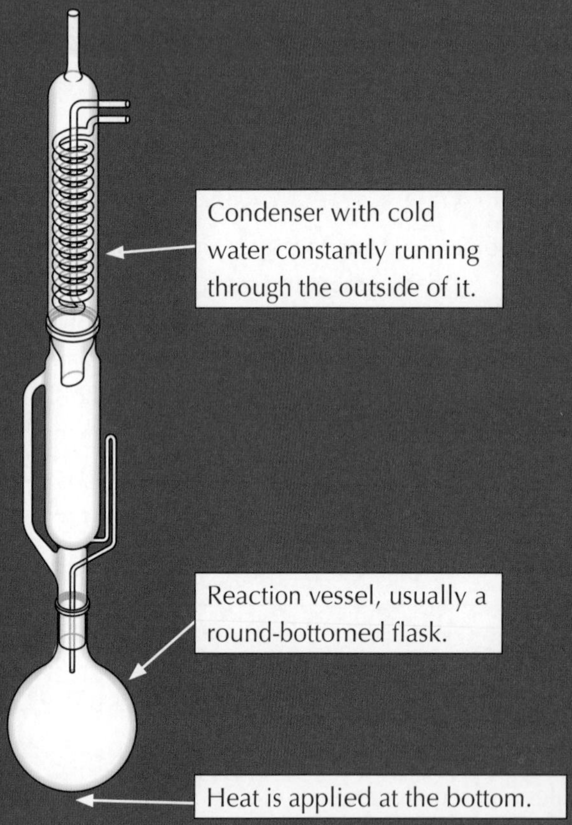

Condenser with cold water constantly running through the outside of it.
Reaction vessel, usually a round-bottomed flask.
Heat is applied at the bottom.

Oxidation

Using cold $K_2Cr_2O_7/H^+$ and a primary alcohol, an aldehyde is formed; this is a new type of functional group with a C double bonded to an O on the end of a chain. The aldehyde will have to be distilled off. (Distillation is the process of boiling off the aldehyde, which will separate it out as it will be heated exactly up to the boiling point of the aldehyde.) For example:

$CH_3CH_2CH_2OH + \text{cold } K_2Cr_2O_7/H^+ \longrightarrow CH_3CH_2COH$, or 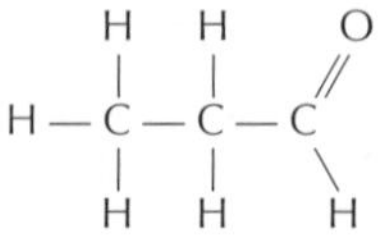

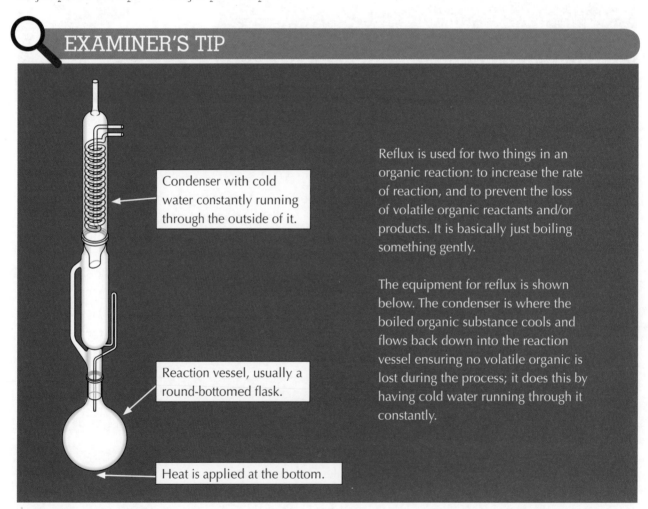

 ISBN: 9780170352611

Using $K_2Cr_2O_7/H^+$, reflux and a primary alcohol gives (after going through the aldehyde and then reacting further) carboxylic acid, a functional group with a C double bonded to an O and an OH. You could also use acidified potassium permanganate here too, which is a stronger oxidant than potassium dichromate. (Note: Potassium permanganate is a stronger oxidant which means the aldehyde will not be able to be isolated out.) For example:

$CH_3CH_2CH_2OH$ + reflux/$K_2Cr_2O_7/H^+ \longrightarrow CH_3CH_2COOH$, or 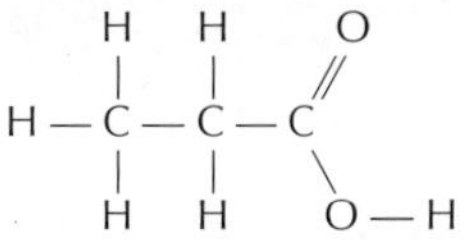

Using $K_2Cr_2O_7/H^+$, reflux and a secondary alcohol, a ketone is formed, a functional group with a C double bonded to an O in the middle of a chain. For example:

$CH_3CH(OH)CH_3$ + reflux/$K_2Cr_2O_7/H^+ \longrightarrow CH_3C(O)CH_3$, or 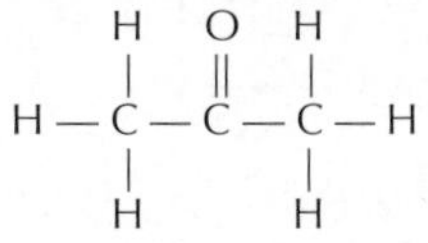

Tertiary alcohols cannot be oxidised.

Elimination

Using 170°C with concentrated H_2SO_4 or Al_2O_3, the water is lost and an alkene is formed. For example:

$CH_3CH(OH)CH_3 + Al_2O_3 \longrightarrow CH_3CHCH_2$

If the molecule is unsymmetrical, then the carbon with the fewest hydrogens will lose another to form the major product. For example:

$CH_3CH(OH)CH_2CH_3$ + c. H_2SO_4/170°C $\longrightarrow$ $CH_3CH{=}CHCH_3$ (major) + $CH_2{=}CHCH_2CH_3$ (minor)

2.3: Alcohols

1 Name or draw the structures of the following alcohols.

Name	Structure
2-methylpropan-2-ol	
	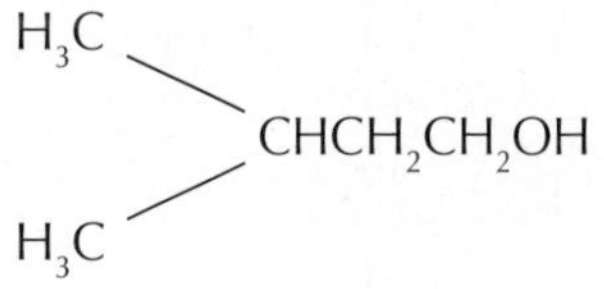
3-chlorobutan-1-ol	

2 Complete these reactions of alcohols.

a $CH_3CH_2CH_2OH$ + cold $H^+/K_2Cr_2O_7 \longrightarrow$ ______

b $CH_3CH_2CH(OH)CH_3$ + reflux/$H^+/K_2Cr_2O_7 \longrightarrow$ ______

c $CH_3CH_2CH_2CH(OH)CH_3$ + 170°C/c. $H_2SO_4 \longrightarrow$ ______

d $CH_3CH_2OH + SOCl_2 \longrightarrow$ ______

e CH_3CH_2OH + reflux/$H^+/K_2Cr_2O_7 \longrightarrow$ ______

f CH_3CH_2OH + cold $H^+/K_2Cr_2O_7 \longrightarrow$ ______

g $CH_3CH(OH)CH_3$ + reflux/$H^+/K_2Cr_2O_7 \longrightarrow$ ______

h $CH_3CH_2OH + Al_2O_3 \longrightarrow$ ______

i $CH_3CH_3CH_2CH_2OH + SOCl_2 \longrightarrow$ ______

j $CH_3CH_3CH_2CH_2OH$ + reflux/$H^+/K_2Cr_2O_7 \longrightarrow$ ______

What do the following key words mean?

Structural isomer	
Geometric isomer	
Optical isomer	
Plane polarised light	
Hydrocarbon	
Alkane	
Alkene	
Alcohol	
Functional group	
Structural formula	
Reflux	
Distillation	

 ISBN: 9780170352611

EXPERIMENT 1

Properties of alcohols

AIM: To determine the differences in properties between primary, secondary and tertiary alcohols.

EQUIPMENT and CHEMICALS:

liquid A
liquid B
liquid C
250 mL beaker
6 x test tubes
potassium dichromate solution (0.1 $molL^{-1}$)
sulfuric acid (2 $molL^{-1}$)
spatula
Bunsen burner

Liquids A, B and C are one of the following chemicals: ethanol, butan-1-ol, 2-methylpropan-2-ol.

SAFETY PRECAUTIONS:

Ethanol is highly flammable and is also an irritant if inhaled or ingested. Butan-1-ol, 2-methylpropan-2-ol and potassium dichromate are irritants to eyes, lungs and skin. Sulfuric acid is corrosive and also an irritant to the skin, eyes and lungs.

METHOD:

Part A — Solubility

Test each of the liquid's solubility by adding the liquid, drop by drop, into 1 mL of water. State if the liquid is soluble, partially soluble, or insoluble.

Part B — Oxidation of alcohols

1 Place 1 mL of each liquid into each of three test tubes.
2 To each liquid add 10 drops of H_2SO_4, then two drops of $K_2Cr_2O_7$, and then warm the solution in a water bath. (This is set up by filling a 250 mL beaker to a quarter full, heating over a Bunsen burner until the water reaches around 60°C, then taking the beaker off the heat and placing the test tube into the beaker.) Record observations and how fast or slow they reacted.

RESULTS:

Alcohol	Solubility	Oxidation
A		
B		
C		

QUESTIONS:

1 Were any of liquids A, B or C soluble in water? If so, what type of intramolecular **and** intermolecular bonds are present?

ISBN: 9780170352611

2 Explain the observed solubility above.

3 Which alcohol(s) are oxidised to aldehydes/carboxylic acids, ketones? Write equations to show the products that are formed.

4 Which liquid is most likely to be the **primary** alcohol? __________

Explain your answer.

5 Which liquid is most likely to be the **tertiary** alcohol? __________

Explain your answer.

 ISBN: 9780170352611

CHECKPOINT 1
WHAT HAVE YOU LEARNED SO FAR?

Isomers, hydrocarbons and alcohols

1 Name or draw the structures of the following molecules.

Name	Structure
2-methylpropan-1-ol	
	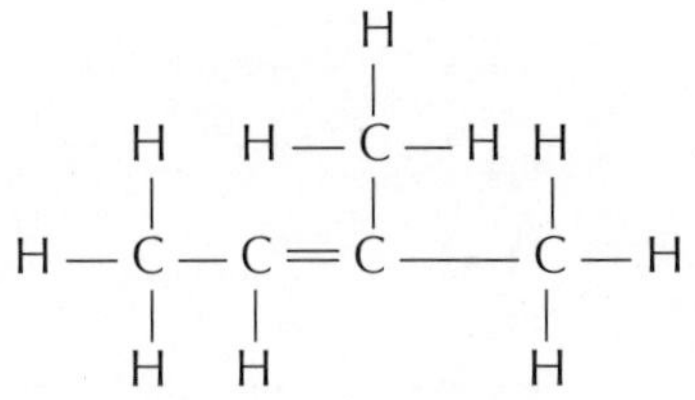
2,2-dimethylhexane	
3-methylhex-1-ene	

2 Complete the reaction scheme shown below.

Name: butane
Structure:

Reagent and conditions:

Name: 1-chlorobutane
Structure:

Reagent and conditions:

Reagent and conditions:

Name: but-2-ene
Structure:

Reagent and conditions:

Name: butan-2-ol
Structure:

Reagent and conditions:

Name: butanone
Structure:

Reflux $K_2Cr_2O_7/H^+$

3 a Which of the following two molecules form optical isomers? Explain your answer.

i

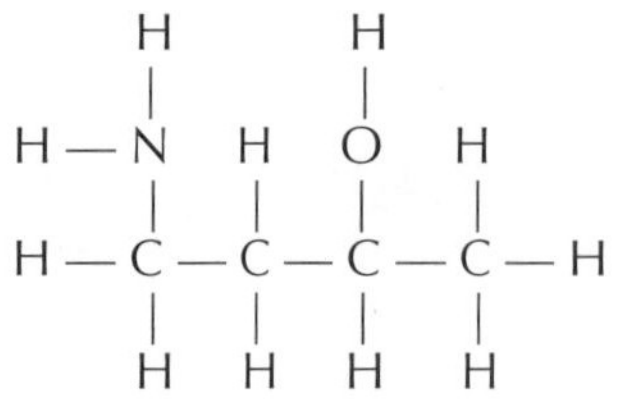

ii

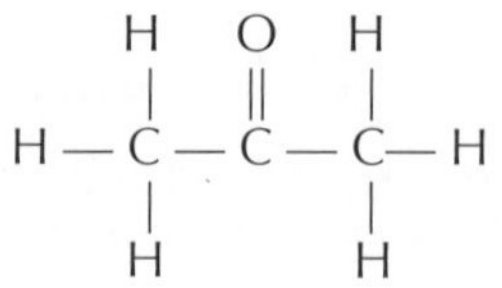

b Draw the two enantiomers in 3D.

c How could you distinguish between the two enantiomers?

4 2,3-dibromobut-2-ene and 2,3-dibromobut-1-ene are structural isomers of each other. Explain why one of them can also form geometric isomers but the other cannot. In your answer include structural formulas and names of the two geometric isomers.

ISBN: 9780170352611

Haloalkanes

Naming

Haloalkanes are named using the same rules as naming an alkane, with the halogen numbered and named in front.
Examples:

$CH_3CH_2CH_2Cl$ is called 1-chloropropane.
$CH_3CH_2CH(Cl)CH_3$ is called 2-bromobutane.

CHEMISTRY APPS

Haloalkanes are used in flame retardants and fire extinguishers, and have many other uses as well.

Physical properties

Haloalkanes contain permanent dipole-dipole bonds in between the molecules, which are caused by the electronegative halogen atom. However, as the chain length increases, the molecules become less and less polar in character, as they have more and more instantaneous dipole-induced dipole bonds.

With up to six carbons, they are generally liquids that easily turn into gases (they are volatile), so they have low melting and boiling points, although these are overall higher than corresponding hydrocarbons.

They are relatively insoluble in water because there is less polar character in them compared with a corresponding alcohol.

Haloalkanes, like alcohols, can be classed as primary, secondary or tertiary:

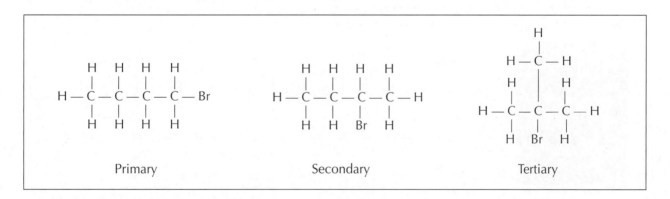

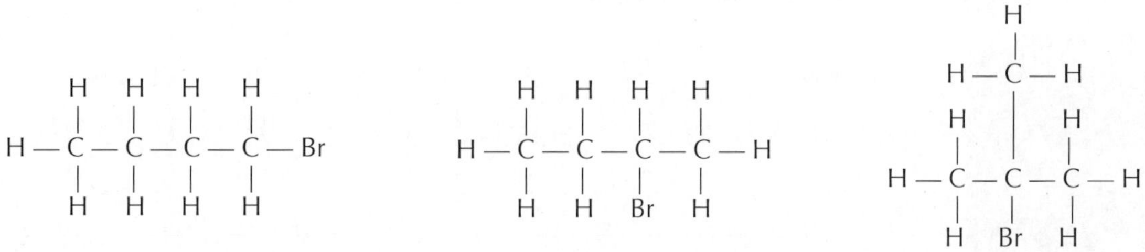

Primary — Secondary — Tertiary

Chemical properties

There are two types of reaction that haloalkanes undergo that you need to know:

Substitution

- Using aqueous NaOH and reflux to produce an alcohol. For example:
  $CH_3CH_2Cl + NaOH_{(aq)}$ reflux $\longrightarrow CH_3CH_2OH$
- Using warm ammonia, NH_3, to produce an amine. For example:
  $CH_3CH_2Cl + NH_3$ warm $\longrightarrow CH_3CH_2NH_2$

Elimination

Favoured when the solvent used is less polar, so we use alcoholic KOH.

This will form an alkene. Note that an unsymmetrical alcohol will produce two products, a major (where the carbon with the least hydrogens will lose another hydrogen) and a minor (where the carbon with the most hydrogens loses one rather than the carbon with the least).

The reaction is favoured with tertiary haloalkanes rather than primary. For example:

$$CH_3C(CH_3)_2(Cl) + KOH_{(alc)} \longrightarrow CH_3C(CH_3)=CH_2$$

2.4: Haloalkanes

1   Name or draw the structures of the following haloalkanes.

Name	Structure
	H—C—H with H—C—C—C—H and Br (2-methyl... bromo structure)
2-bromo-3-chlorobutane	
	$CH_3CH(Cl)CH(CH_3)_2$

2   Label the following haloalkanes as primary, secondary or tertiary.

Structure	$CH_3 — \overset{\overset{CH_3}{|}}{\underset{\underset{Cl}{|}}{C}} — CH_2CH_3$	$CH_3 — \overset{\overset{CH_3}{|}}{\underset{\underset{Br}{|}}{C}} — CH_3$	$H — \overset{\overset{Cl}{|}}{\underset{\underset{H}{|}}{C}} — \overset{\overset{CH_3}{|}}{\underset{\underset{CH_3}{|}}{C}} — CH_3$
Classification			

3   Complete the following reactions, which either prepare or react haloalkanes.

a $CH_3CH_2CH_3 + Cl_2 \xrightarrow{UV}$ _____

b $CH_3CH=CH_2 + HCl\ (CCl_4) \longrightarrow$ _____

c $CH_3CH_2CH_2Br + NH_3\ (warm) \longrightarrow$ _____

d $CH_3CH(CH_3)CH_2Br + KOH_{(aq)} \xrightarrow{reflux}$ _____

e $CH_3C(Br)(CH_3)_2 + KOH_{(alc)} \xrightarrow{reflux}$ _____

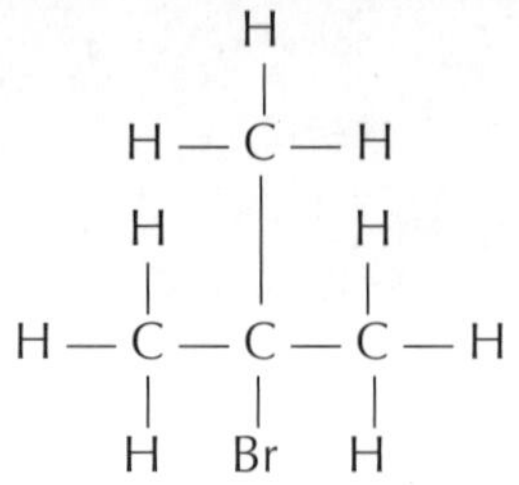

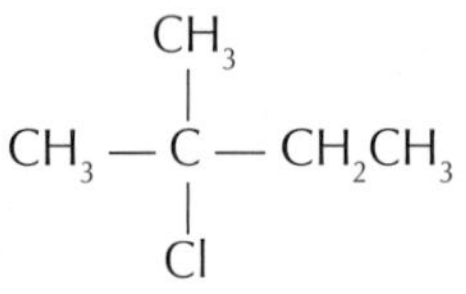

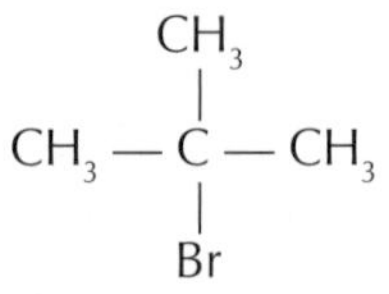

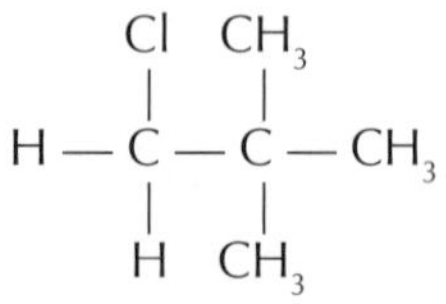

Amines

Naming

There are two ways to name an amine: add either the prefix 'amino' or the suffix 'amine'; either way it needs to be numbered in order to know which carbon number the NH_2 group comes off.

Example:
$CH_3CH_2CH_2NH_2$ is called 1-aminopropane or propan-1-amine.

CHEMISTRY APPS

Many neurotransmitters are amines. A neurotransmitter is a chemical substance which is released at the end of a nerve fibre because of a nerve impulse.

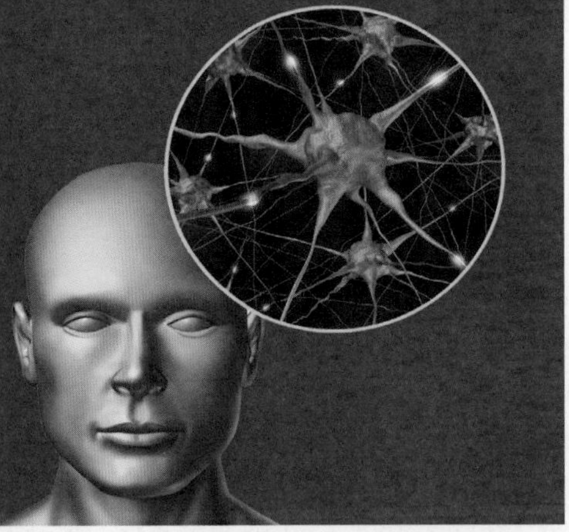

Physical properties

Amines contain hydrogen bonds between their molecules, which gives them higher melting and boiling points than corresponding haloalkanes but lower than alcohols. This also allows them to be soluble in water too if the chain length is relatively short (less than five carbons long). Amines have fishy smells. They are basic and that is why we put lemon or vinegar (acids) on fish in order to neutralise the smell.

Amines can also be classified as primary, secondary or tertiary, depending on how many carbons are attached to the carbon attached to the N of the amine group or depending on how many carbons are attached to the N:

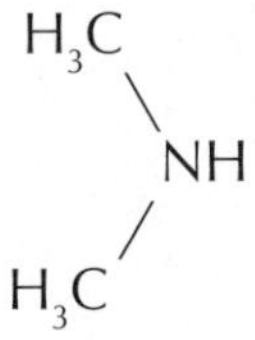

N-dimethylamine (secondary)

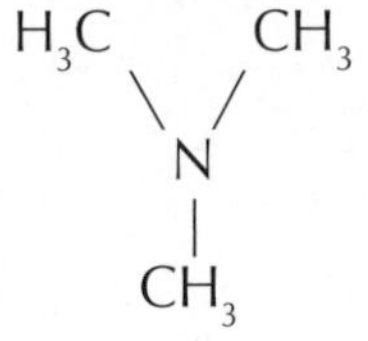

N,N-trimethylamine (tertiary)

CHEMISTRY APPS

The active ingredient in asthma inhalers is salbutamol, which contains an amine functional group.

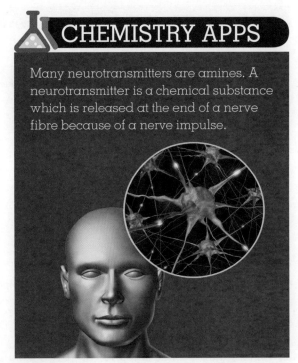

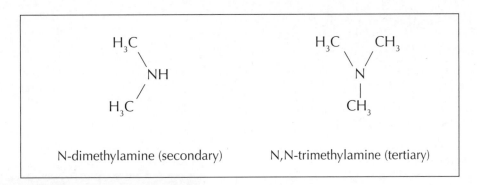

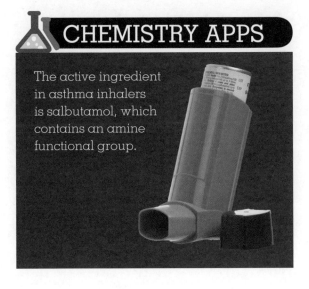

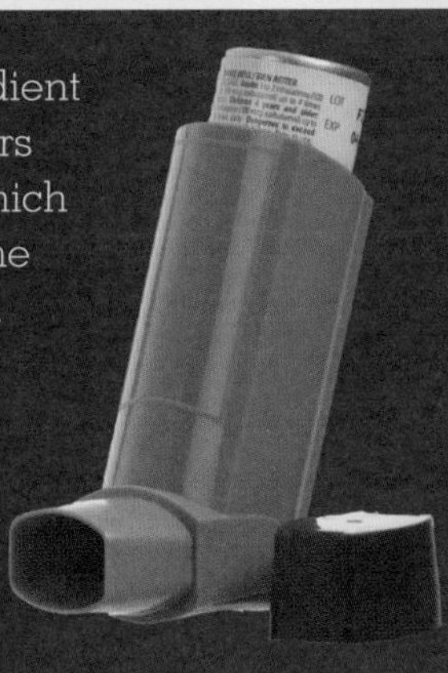

Chemical properties

As amines are basic they will undergo the following reactions.

- Water-soluble amines form alkaline solutions. For example:
  $CH_3NH_2 + H_2O \longrightarrow CH_3NH_3^+ + OH^-$
- Amines also react with acids to form salts. For example:
  $CH_3NH_2 + HCl \longrightarrow CH_3NH_3^+Cl^-$
  aminomethane   methylammonium chloride

ISBN: 9780170352611

2.5: Amines

1 Name or draw the following amines:

Name	Structure
	$CH_3CH_2CH(CH_3)CH_2NH_2$
methanamine	
	$$\begin{array}{c} CH_3 \\ | \\ H_3C - CH - CH_2 - NH_2 \end{array}$$

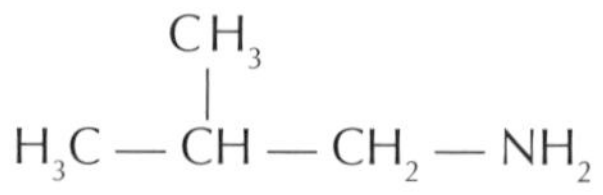

2 Classify the following amines as primary, secondary or tertiary.

Structure	Classification
$$\begin{array}{ccccc} H & H & H & H & H \\ | & | & | & | & | \\ H - C - C - N - C - C - H \\ | & | & \cdot\cdot & | & | \\ H & H & & H & H \end{array}$$	
$$\begin{array}{c} CH_3 \\ | \\ H_3C - N - CH_2 - CH_3 \end{array}$$	
$CH_3CH_2CH(NH_2)CH_3$	
CH_3NH_2	

3 Complete the following reactions of amines.

a $CH_3CH_2NH_2 + HNO_3 \longrightarrow$ _____

b $CH_3CH_2NH_2 + H_2O \rightleftharpoons$ _____

c $CH_3CH_2Br + NH_3$ (warm) $\longrightarrow$ _____

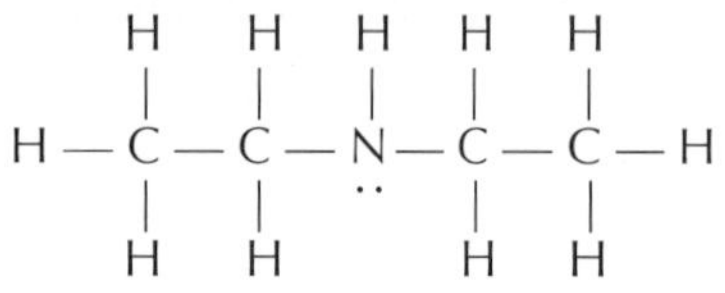

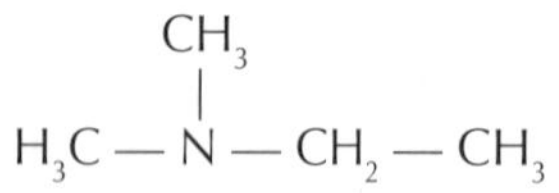

EXPERIMENT 2

Amines

AIM: To observe the properties of amines.

EQUIPMENT and CHEMICALS:

aminopropane
universal indicator
red and blue litmus paper
test tubes

SAFETY PRECAUTIONS:

Aminopropane is an irritant if inhaled or ingested. Universal indicator is an irritant to the lungs, as well as to skin.

METHOD:

1 ***Carefully*** smell the amine by using the wafting technique (waving your hand over the top of it).

2 Test the amine with damp red and blue litmus and universal indicator.

RESULTS:

	Amine
Smell	
Litmus	
Universal indicator	

QUESTIONS:

1 Write an equation to show why amines are basic in aqueous solution.

2 Would you expect a longer chain or a shorter chain amine to be more soluble in water? Why?

ISBN: 9780170352611

Aldehydes and ketones

Naming

Aldehydes are named with the suffix 'al'; they have the functional group $C(=O)-H$, which does not need to be numbered as it can only occur on the end of the molecule.

Example:

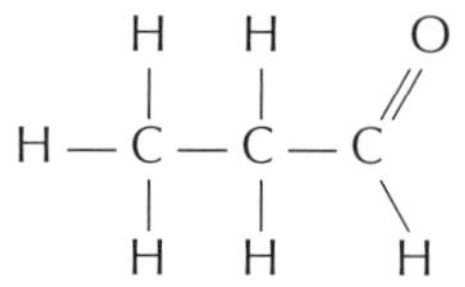

is called propanal.

Ketones have the suffix 'one'; the functional group does need to be numbered because the carbonyl group (C=O) can occur anywhere in the middle of the molecule.

Example:

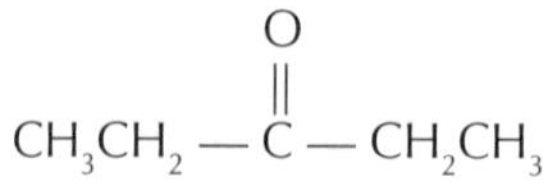

$$CH_3CH_2-\overset{\overset{\large O}{\|}}{C}-CH_2CH_3$$ is called pentan-3-one.

CHEMISTRY APPS

Propanone or acetone is used in nail polish remover.

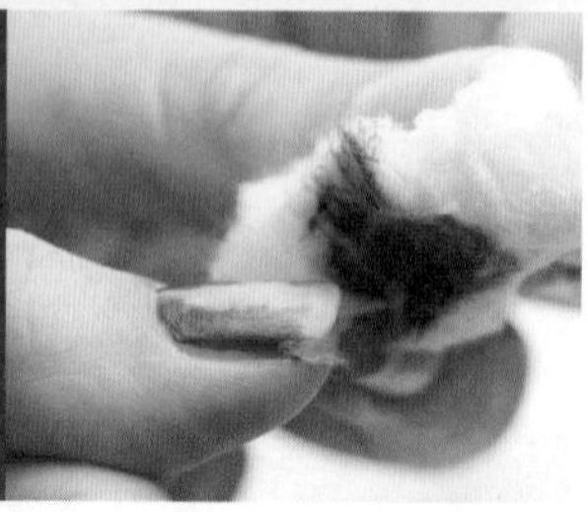

Physical properties

Most are volatile liquids that have definite odours, with aldehydes generally smelling worse than ketones. The lone pair of electrons on the oxygen means that there is the possibility for hydrogen bonding with water, however there is none between their molecules. This allows them some solubility in water and slightly higher melting and boiling points than their corresponding alkanes. However, as chain length increases, their solubility in polar solvents also decreases.

Chemical properties

Ketones are generally quite unreactive, but aldehydes readily oxidise to carboxylic acids using very mild oxidants like Tollens' reagent, Fehling's solution or Benedict's solution. See the details shown below.

Oxidising agent	Colour change
Tollens' reagent — ammoniacal silver nitrate, or $[Ag(NH_3)_2]^+$	Colourless solution forms silver precipitate or silver mirror
Fehling's solution — Cu^{2+} in basic conditions, forms Cu_2O	Blue solution forms orange-red precipitate
Benedict's solution — Cu^{2+} in basic conditions, forms Cu_2O	Blue solution forms orange-red precipitate

You can also reduce an aldehyde or ketone back into an alcohol using sodium borohydrate, $NaBH_4$.

 ISBN: 9780170352611

2.6: Aldehydes and ketones

1 Name or draw the structures of the following molecules.

Name	Structure
Pentan-2-one	
Pentanal	
	$CH_3CH_2C(O)CH_3$
	CH_3CH_2COH

2 Complete the following reactions.

a CH_3CH_2COH + Benedict's solution $\longrightarrow$

b $CH_3CH(CH_3)COH$ + Fehling's solution $\longrightarrow$

c CH_3COH + Tollens' reagent $\longrightarrow$

d CH_3CH_2OH + cold $K_2Cr_2O_7/H^+$ $\longrightarrow$

e $CH_3CH(OH)CH_3$ + reflux $K_2Cr_2O_7/H^+$ $\longrightarrow$

3 Explain a test that would help you distinguish between samples of butanone and butanal; include any relevant observations and equations.

ISBN: 9780170352611

EXPERIMENT 3

Oxidation of carbonyl compounds

AIM: To compare the different oxidants that can be used to oxidise aldehydes.

EQUIPMENT and CHEMICALS:

potassium dichromate solution (0.1 $molL^{-1}$)
sodium hydroxide solution (2 $molL^{-1}$)
sulfuric acid (2 $molL^{-1}$)
silver nitrate solution (0.01 $molL^{-1}$)
Bunsen burner
propanal
propanone
glucose
thermometer
tripod and gauze
ammonia solution (2 $molL^{-1}$)
Benedict's solution
250 mL beaker
9 test tubes
test tube holders

SAFETY PRECAUTIONS:

Potassium dichromate, sodium hydroxide, sulfuric acid and propanal are skin, eye and lung irritants; as is ammonia but it is also flammable. Propanone is an eye and lung irritant. Sulfuric acid is also corrosive.

METHOD:

Carry out each of the tests on propanal, propanone and glucose.
Record your result in the table.

Test 1 — Oxidation

1. Add 1 mL of sample (either propanal, propanone or glucose) to 1 mL of potassium dichromate and 1 mL of sulfuric acid.
2. Warm in a water bath (set up in the 250 mL beaker by filling about a quarter full with water and then heating to around 60°C on a Bunsen burner and then setting the beaker aside and placing your test tube inside).

Test 2 — Tollens' reagent test

1. Place 5 mL of aqueous silver nitrate (0.01 $molL^{-1}$) in three clean test tubes and add one drop of 2 $molL^{-1}$ sodium hydroxide solution.
2. Add 2 $molL^{-1}$ ammonia solution until the precipitate just dissolves.
3. Add five drops of propanal to one test tube, five drops of propanone to another test tube, and five drops of glucose to the third test tube.
4. Warm in a water bath below 60°C for 10 minutes (see instructions above for setting this up).

Test 3 — Oxidation

1. Add 1 mL of Benedict's solution to 2 mL of each solution.
2. Heat ***gently*** over a Bunsen burner. (Heating gently means wave it in and out of a blue flame; remember to point the top of the test tube towards the wall and away from people.)

RESULTS:

	Observations		
	Test 1	Test 2	Test 3
Propanal			
Propanone			
Glucose			

ISBN: 9780170352611

QUESTIONS:

1 Why does propanal and glucose react in similar ways?

2 Write an equation for the reaction for propanal and Tollens' reagent.

3 Write an equation for the reaction for propanal and Benedict's solution.

What do the following key words mean?

Halolalkane	
Amine	
Substitution	
Elimination	
Oxidation	
Aldehyde	
Ketone	
Carbonyl	

Names, properties and reactions of haloalkanes, amines, aldehydes and ketones

1 Name or draw the structures for the following molecules.

Name	Structure
2-methylbutan-1-amine	
	$CH_3CH_2CH(CH_3)CH_2Br$
	$CH_3CH_2CH_2COH$
hexan-2-one	

2 Explain how you could distinguish between aminoethane, ethanal and propanone using damp red litmus and Benedict's solution. Include any equations in your answer.

EXAMINER'S TIP

Litmus must always be damp in order to show a positive result.

 ISBN: 9780170352611

CHECKPOINT 2
WHAT HAVE YOU LEARNED SO FAR?

3 Complete this reaction scheme.

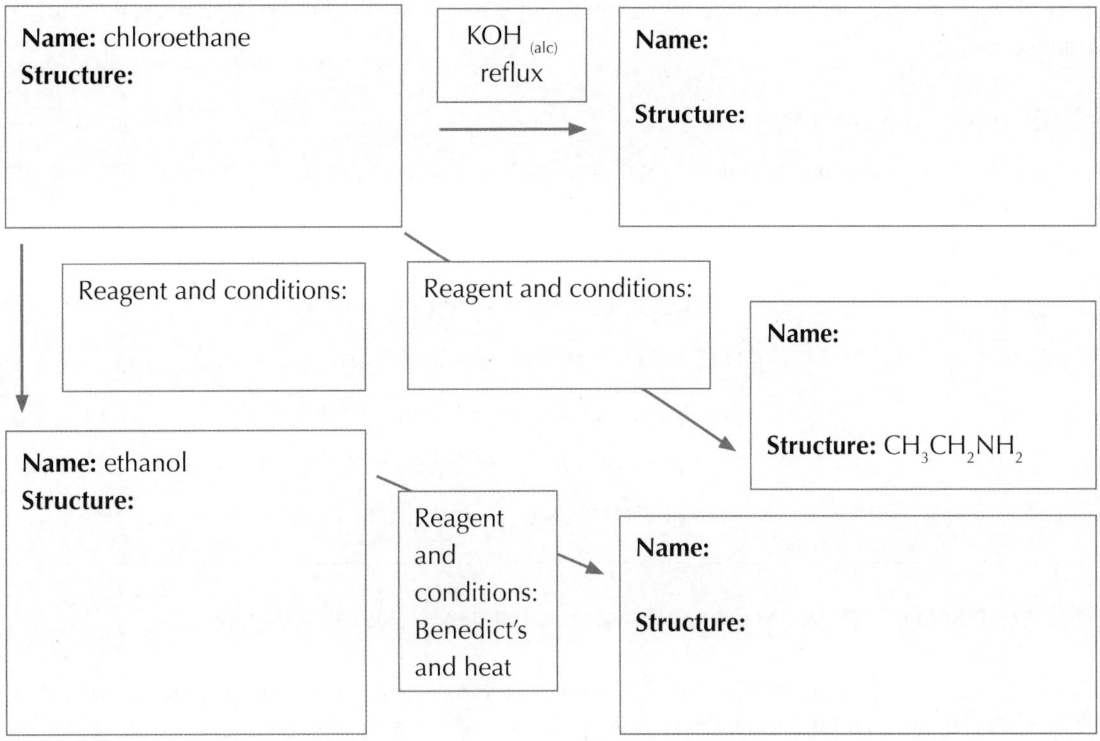

Carboxylic acids and acid chlorides

Naming

Carboxylic acids are named with the suffix 'oic acid'; they have the functional group COOH (with one O being double bonded).
Example:

$CH_3CH_2CH_2COOH$ is called butanoic acid.

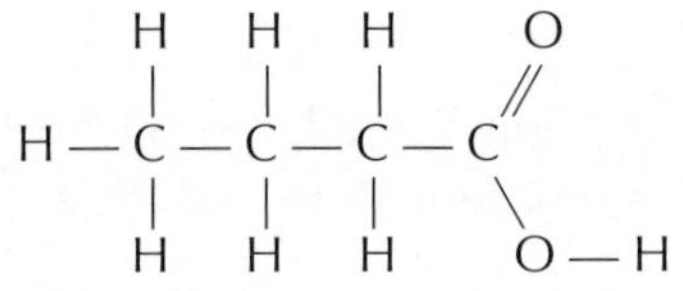

FACT

The original names for organic acids were from the natural compounds found within them. So methanoic acid was called formic acid, as it was found in red ants (*formica* in Latin means ant).

Acid (or acyl) chlorides are named with the suffic 'oyl chloride'; they have the functional group COCl (also with the O being double bonded).
Example:

CH_3CH_2COCl is called propanoyl chloride.

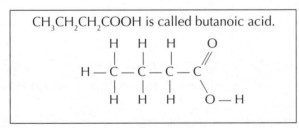

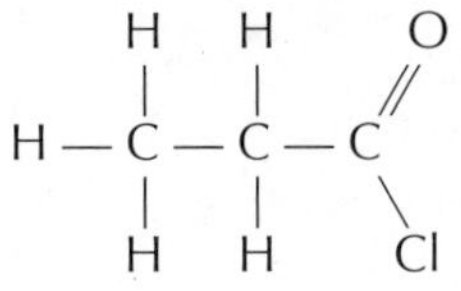

Physical properties

Carboxylic acids and acid chlorides are both acids with pungent smells.

Carboxylic acids have hydrogen bonds so are soluble in water when the chain length is small and this decreases as the chain length increases. These hydrogen bonds also allow carboxylic acids to have high melting and boiling points.

ISBN: 9780170352611

Acid chlorides have a strong polar end due to the O and the Cl, giving them permanent dipole-dipole bonds. This gives them some ability to dissolve in water but overall they are less soluble in water than carboxylic acids. They have high melting and boiling points but lower than corresponding carboxylic acids. They are very volatile.

Chemical properties

Carboxylic acids are weak acids and so will dissociate in water to produce hydronium ions, neutralise bases and make metal salts and hydrogen gas. For example:

$$CH_3COOH + H_2O \rightleftharpoons CH_3COO^- + H_3O^+$$

$$CH_3COOH + NaOH \longrightarrow CH_3COO^- + Na^+ + H_2O$$

$$ZnO + 2CH_3COOH \longrightarrow Zn(CH_3COO)_2 + H_2O$$

$$Na_2CO_3 + 2CH_3COOH \longrightarrow 2NaCH_3COO + CO_2 + H_2O$$

$$2CH_3COOH + Mg \longrightarrow Mg(CH_3COO)_2 + H_2$$

Their reaction rates with bases and metals are slow but they will go to completion.

Carboxylic acids will also undergo three different reactions (shown below) to produce acid chlorides, esters (R – COO – R′) and amides ($R – CONH_2$).

- To prepare an acid chloride, the reagent is thionyl chloride, $SOCl_2$:

  $CH_3COOH + SOCl_2 \xrightarrow{\text{heat}} CH_3COCl$

  ethanoic acid → ethanoyl chloride

- To prepare an ester, an alcohol and concentrated sulfuric acid, H_2SO_4, are required:

  $CH_3COOH + CH_3OH \xrightarrow{\text{heat c. } H_2SO_4} CH_3COOCH_3$

  ethanoic acid + methanol $\xrightarrow{\text{heat c. } H_2SO_4}$ methyl ethanoate

  The OH is lost from the carboxylic acid and the H is lost from the alcohol; this is where the two add on to each other to form the ester and water. This is done under reflux conditions and a base is added at the end of the process to neutralise any remaining sulfuric acid.

- To prepare an amide, add aqueous ammonia, NH_3, and heat, or if you add a primary or a secondary amine, you can make a secondary or a tertiary amide respectively:

  $CH_3COOH + NH_3 \longrightarrow CH_3COO^-NH_4^+ \xrightarrow{\text{heat}} CH_3CONH_2$

  ethanamide

  $CH_3COOH + CH_3NH_2 \longrightarrow CH_3COO^-CH_3NH_3^+ \xrightarrow{\text{heat}} CH_3CONHCH_3$

  N-methylethanamide

  $CH_3COOH + CH_3NHCH_2CH_2CH_3 \longrightarrow CH_3COO^-CH_3NH_2^+CH_2CH_3 \xrightarrow{\text{heat}} CH_3CON(CH_3)CH_2CH_3$

  N, N-ethylmethylethanamide

ISBN: 9780170352611

Acid chlorides react rapidly with water to produce a carboxylic acid and white fumes of hydrochloric acid making a very acidic solution:

$$CH_3COCl + H_2O \longrightarrow CH_3COOH + HCl$$

By adding an alcohol they will also form esters:

$$CH_3COCl + CH_3CH_2CH_2OH \longrightarrow CH_3COCH_2CH_2CH_3$$

propyl ethanoate

2.7: Carboxylic acids and acid chlorides

1 Name or draw the following structures.

Name	Structure
2-methylbutanoic acid	
3-chloro-2,3-dimethylpentanoic acid	
	$CH_3CH(CH_3)CH_2COCl$
butanoyl chloride	

2 Complete the following reactions.

a $CH_3CH_2OH + H^+/K_2Cr_2O_7 \xrightarrow{\text{reflux}}$

b $CH_3CH(CH_3)CH_2COOH + SOCl_2 \longrightarrow$

c $CH_3CH_2CH(CH_3)COOH + NaOH \longrightarrow$

d $CH_3CH_2COOH + CaCO_3 \longrightarrow$

e $CH_3COOH + CH_3CH_2CH_2OH \xrightarrow{\text{c. } H_2SO_4}$

f $CH_3CH_2COCl + CH_3OH \longrightarrow$

g $CH_3CH_2COOH + NH_{3\,(aq)} \xrightarrow{\text{heat}}$

h $CH_3COCl + H_2O \longrightarrow$

i $CH_3CH(CH_3)COOH + H_2O \rightleftharpoons$

3 Explain how you could identify samples of ethanoic acid and ethanoyl chloride using water and damp blue litmus. Include any relevant observations and equations.

Esters and amides

Naming

Esters are named for the section that is part of the alcohol (methyl if it was from methanol, ethyl if it was ethanol, …) and then the carboxylic acid/acid chloride part (methanoate if it is from methanoic acid or methanoyl chloride, …). They have the functional group C double bond O and single bond O.

Example:

$CH_3CH_2CH_2COOCH_2CH_2CH_3$ is called propyl butanoate.

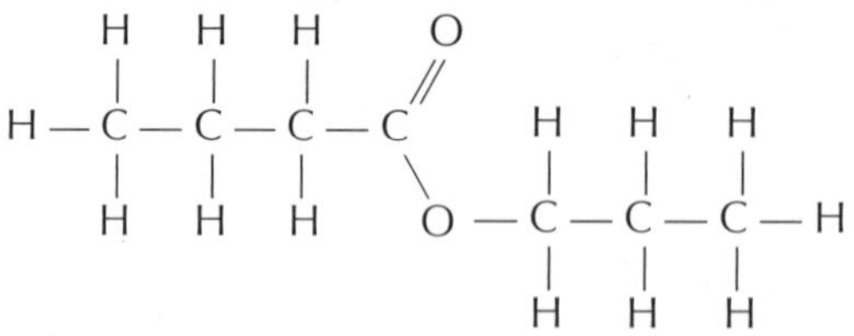

CHEMISTRY APPS

Esters are often used in perfumes because of their nice smells.

Amides are named with the suffix 'amide' and have the functional group C double bond O and an NH_2.

Example:

$CH_3CH_2CONH_2$ is called propanamide.

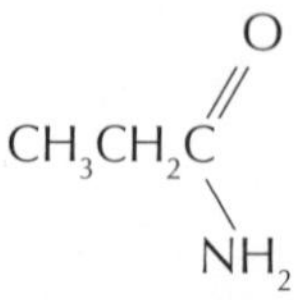

 ISBN: 9780170352611

They can also form secondary and tertiary amides with more than one alkyl chain coming off the N, but you do not need to name these.

Physical properties

Esters are generally not very soluble in water (except for methyl methanoate) because they do not contain hydrogen bonds, but they do contain permanent dipole-dipole bonds. They are colourless, volatile liquids that have fruity smells. They have high melting and boiling points due to these permanent dipole-dipole bonds but lower than alcohols.

Amides are all white solids (except for methanamide) so have high melting and boiling points due to the hydrogen bonds between the O and the H connected to the N of corresponding molecules. They are not basic like amines. Small amides tend to be soluble in water but as the chain length increases, their solubility in water decreases.

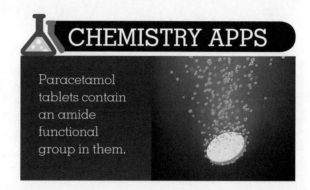

CHEMISTRY APPS

Paracetamol tablets contain an amide functional group in them.

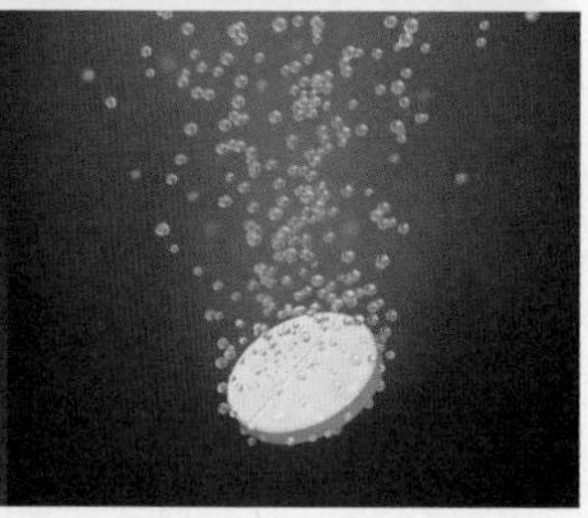

Chemical properties

Esters can be hydrolysed under two different sets of conditions:

- **Acidic** — Using dilute H^+ to form the carboxylic acid and alcohol they were made from.
  $CH_3COOCH_2CH_3 + \text{dil. } H^+ \longrightarrow CH_3COOH + CH_3CH_2OH$

- **Basic** — Using OH^- to form a salt and the alcohol or using ammonia (alc) to form an amine and the alcohol.
  $CH_3COOCH_2CH_3 + OH^- \longrightarrow CH_3CH_2OH + CH_3COO^-$
  $CH_3COOCH_2CH_3 + NH_{3\,(alc)} \longrightarrow CH_3CH_2OH + CH_3CONH_2$

A special hydrolysis of a triester can be used to make a soap and glycerol (propan-1, 2, 3, -triol). This reaction is known as **saponification**:

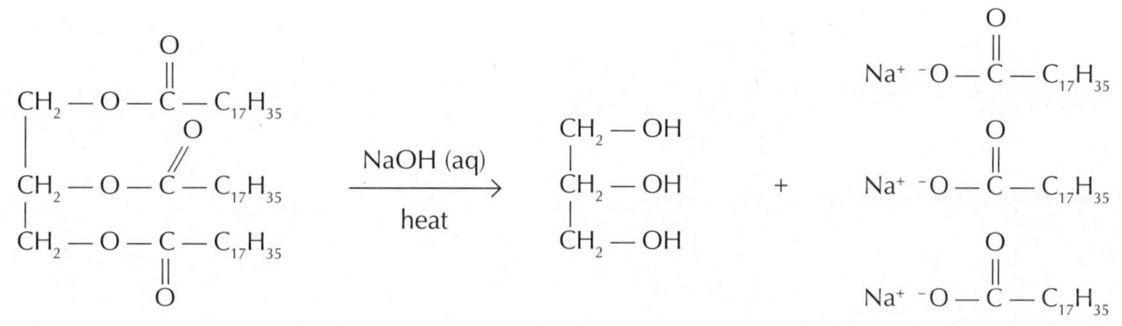

Tristearin, an animal fat — Glycerol — Sodium stearate, a soap

Amides can also be hydrolysed using acidic or basic conditions:

- **Acidic** conditions, which form a carboxylic acid and NH_4^+, ammonium.
  $CH_3CONH_2 + H_3O^+ \longrightarrow CH_3COOH + NH_4^+$

- **Basic** conditions, which form a salt and NH_3, ammonia.
  $CH_3CONH_2 + OH^- \longrightarrow CH_3COO^- + NH_3$

2.8: Amides and esters

1 Name or draw the structure of the following molecules.

Name	Structure
	CH_3COOCH_3
	$(CH_3)_3CCONH_2$
propanamide	
propyl butanoate	

2 Complete the following reactions.

a $CH_3CH_2COOH + CH_3OH \xrightarrow{\text{heat c. } H_2SO_4}$

b $CH_3COCl + CH_3CH_2CH_2OH \longrightarrow$

c $CH_3CH_2COOH + NH_{3\,(aq)} \xrightarrow{\text{heat}}$

d $CH_3CH(CH_3)COOCH_3$ + dil. $H^+ \longrightarrow$

e $CH_3CH(CH_3)COOCH_3 + NH_{3\,(alc)} \longrightarrow$

f $CH_3CH(CH_3)COOCH_3 + OH^- \longrightarrow$

g $CH_3CH(CH_3)CONH_2 + H_3O^+ \longrightarrow$

h $CH_3CH(CH_3)CONH_2 + OH^- \longrightarrow$

ISBN: 9780170352611

EXPERIMENT 4

Preparation of esters

AIM: To prepare esters.

EQUIPMENT and CHEMICALS:

ethanol
ethanoic acid
pentan-1-ol
3-methylbutan-1-ol
methanoic acid
octan-1-ol
methanol
salicylic acid
sodium carbonate
kettle
6 x test tubes
250 mL beaker
stopwatch
conc. sulfuric acid

SAFETY PRECAUTIONS:

All the acids and sodium carbonate are corrosive and irritants; all the alcohols are flammable. The concentrated sulfuric acid should only be used in a fumehood.

METHOD:

1 Prepare a water bath in a 250 mL beaker (by quarter filling the beaker with water and heating over a Bunsen flame until it reaches around 80°C and then setting aside or by pouring boiling water from the kettle into the beaker).
2 Add 1 mL of each alcohol to 1 mL of each carboxylic acid with one drop of concentrated sulfuric acid and place in a hot water bath for two minutes.
3 Add in a little sodium carbonate solution in a small beaker.
4 Record your observations and name the ester in the table below.

Alcohol	Carboxylic acid	Observation/odour	Name of ester
Ethanol	Ethanoic acid		**a**
Pentan-1-ol	Ethanoic acid		**b**
Ethanol	Methanoic acid		**c**
Octan-1-ol	Ethanoic acid		**d**
Methanol	Salicylic acid		**e**
3-methylbutan-1-ol	Ethanoic acid		**f**

QUESTIONS:

1 The sulfuric acid is added for two reasons. What are they?

2 What is the sodium carbonate added for?

3 Write balanced equations for the six esters you made.

a

b

c

d

e

f

Polymers

A polymer is a giant covalent molecule made of simple repeating units known as monomers joined together.

The properties of a polymer depend on:

- the type of bond between the monomers
- what organic functional group the monomers are made from.

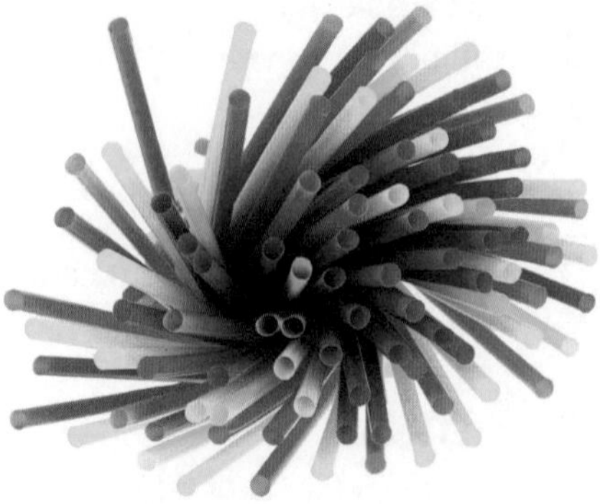

There are two main types of polymer:

- addition, which we saw earlier in the chapter under the hydrocarbons section
- condensation.

Condensation polymers

Here, elimination of a simple molecule joins the monomers, which have two functional groups on them. Monomers commonly have either carboxylic acid, alcohol, amide, amine, ester or acid chloride functional groups and come together to form ester (– COO –) or amide (– CONH –) linkages. Common examples include *polyesters*, *polyamides* and *proteins*.

CHEMISTRY APPS

Most of your school uniform is probably made from terylene or polyester because it is relatively cheap, hard wearing and washable.

Example 1: terylene

ethan-1,2-diol: $HOCH_2CH_2OH$ + benzene-1,4-dicarboxylic acid:
$HOOC - C_6H_4 - COOH \longrightarrow O - CH_2 - CH_2 - OOC - C_6H_4 - CO$

Example 2: nylon

Nylon is one of the original polyamides used to make faux silk. Now it has a wide variety of uses, such as pipes and stockings.

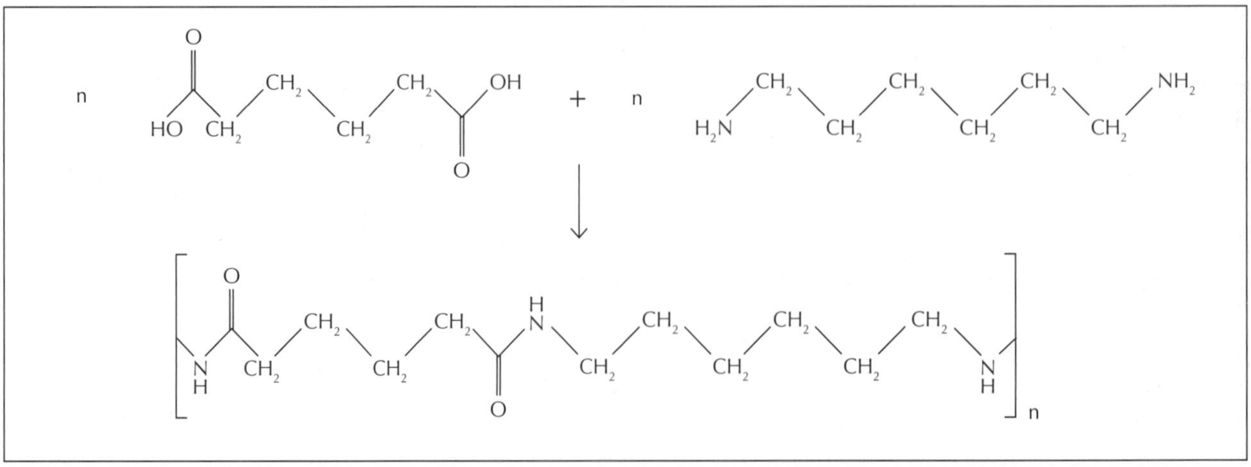

 ISBN: 9780170352611

Proteins

Proteins are made from amide/peptide linkages between amino acids. Amino acids contain two functional groups:

- an amino group NH_2
- a carboxylic acid group COOH.

An amino acid has both an acidic and a basic functional group and so it will commonly exist as a ***zwitterion***. Because they donate and accept protons to form these ions, this means amino acids are very soluble in water and they form solids with high melting points.

Many amino acids joined together are known as a polypeptide. When a polypeptide has a specific function, it is known as a protein, for example insulin, collagen, an enzyme.

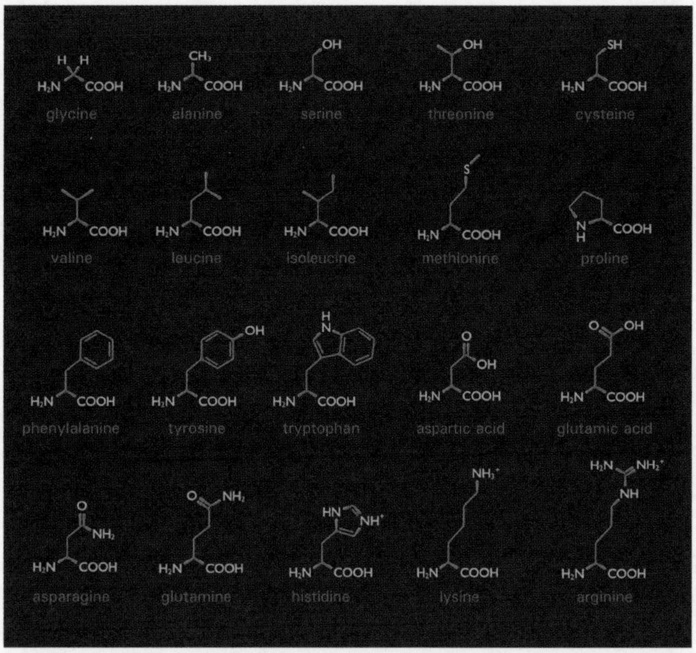

The drawings above are in skeletal form where each corner represents a carbon. The hydrogens are present but omitted in order to simplify the structure.

Note: The amino acids drawn above are drawn in skeletal structural formulas where each corner is a carbon atom and the hydrogens are not drawn at all.

Proteins have several levels of organisation:

- The ***primary structure*** is the basic amino acid sequence.
- The ***secondary structure*** is the 3D arrangement caused by H-bonding from a carboxyl acid O to an NH_2 H.
- The ***tertiary structure*** is the folding of the whole molecule and further interactions between groups.

FACT

The term 'protein' was first used in 1838 and is derived from the Greek word *proteios*, meaning the first rank of importance. Twenty per cent of your body mass is protein.

2.9: Polymers and proteins

1 Complete the following polymer reactions by drawing at least two repeating units.

a $HO - CH_2 - OH + HOOC - CH_2CH_2 - COOH \longrightarrow$

b $HOOC - (CH_2)_{16} - COOH + H_2N - CH_2 - NH_2 \longrightarrow$

c $HOOC - (CH_2)_{12} - CH_2OH + HOOC - CH_2 - OH \longrightarrow$

2 Draw the possible monomers from the polymers shown below.

a

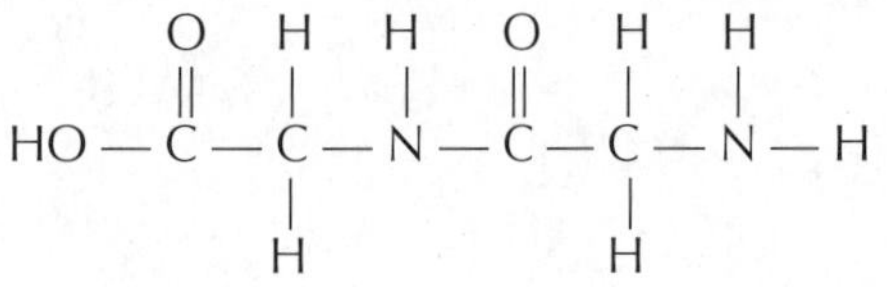

b

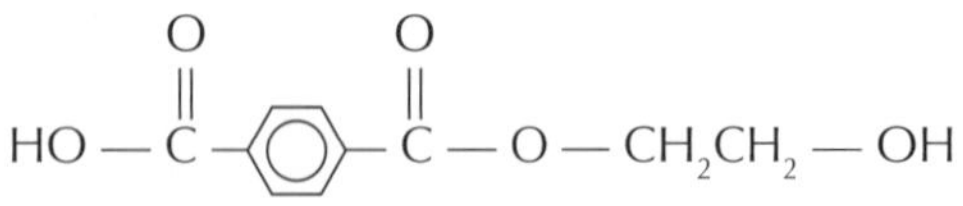

c

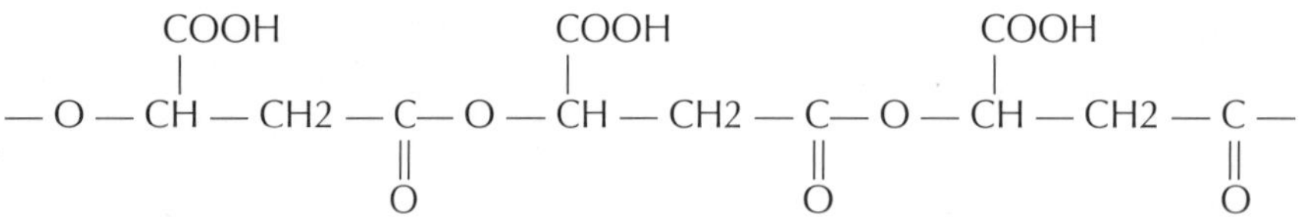

3 Draw the following dipeptide structures by joining the amino acids together below.

a

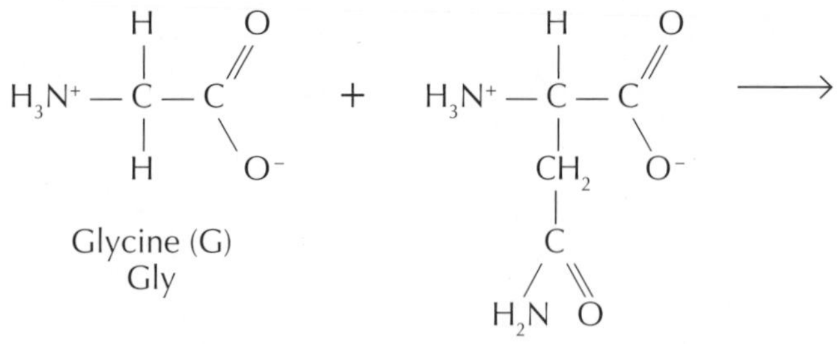

Glycine (G)
Gly

Asparagine (N)
Asn

ISBN: 9780170352611

b

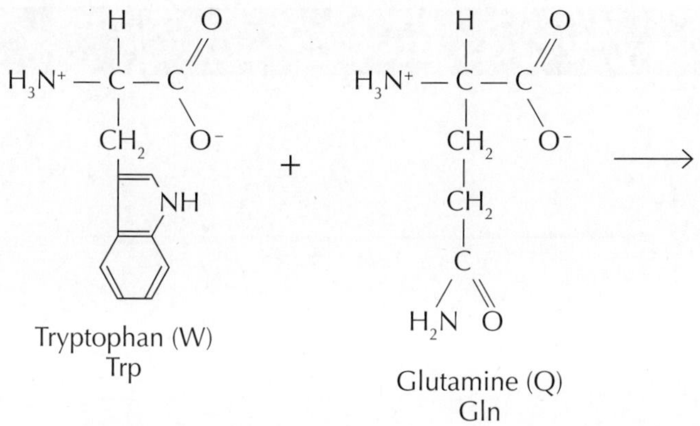

Tryptophan (W) Trp

Glutamine (Q) Gln

c

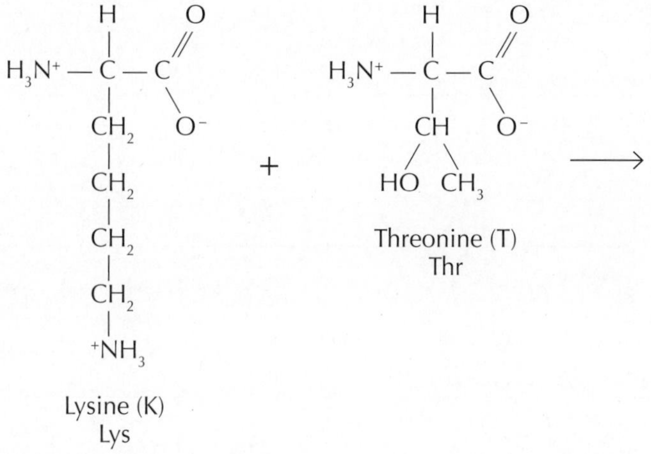

Lysine (K) Lys

Threonine (T) Thr

4 Explain why the formation of all these polymers is considered to be a condensation reaction.

What do the following key words mean?

Carboxylic acid	
Acid chloride	
Ester	
Amide	
Addition polymer	
Condensation polymer	
Amino acid	
Protein	
Monomer	
Zwitterion	
Saponification	

ISBN: 9780170352611

CHECKPOINT 3
WHAT HAVE YOU LEARNED SO FAR?

Names, properties and reactions of carboxylic acids, acid chlorides, amides and esters

1 Name or draw the structures of the following organic molecules.

Name	Structure
2-methylpropanoic acid	
	$CH_3CH(CH_3)CH_2CONH_2$
3-methylbutanoyl chloride	
	$CH_3CH_2COOCH_3$

2 Explain how you could distinguish between samples of propanoic acid, propanoyl chloride and propyl methanoate using water and damp blue litmus paper. Include any relevant observations and equations.

3 Explain why the boiling point of propyl methanoate is much lower than the boiling point of butanamide despite having the same number of carbon atoms in their chains.

4 Complete this reaction scheme.

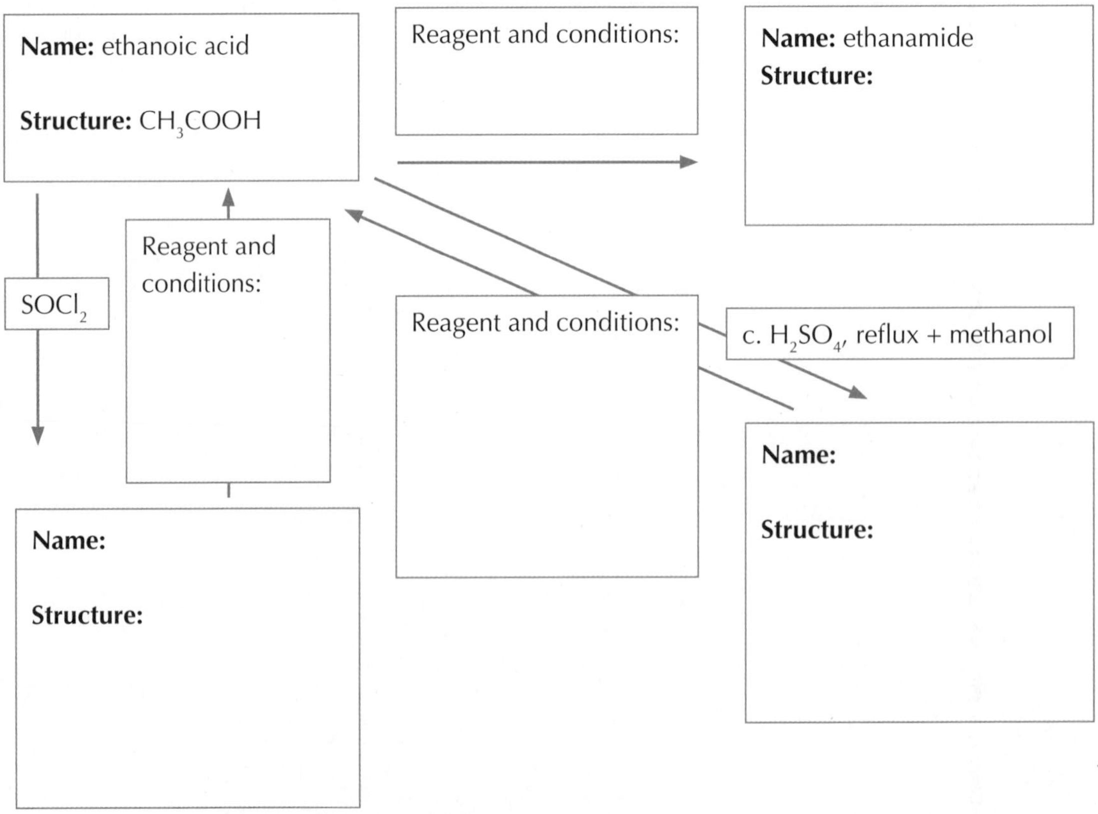

  ISBN: 9780170352611

Organic reaction scheme summary

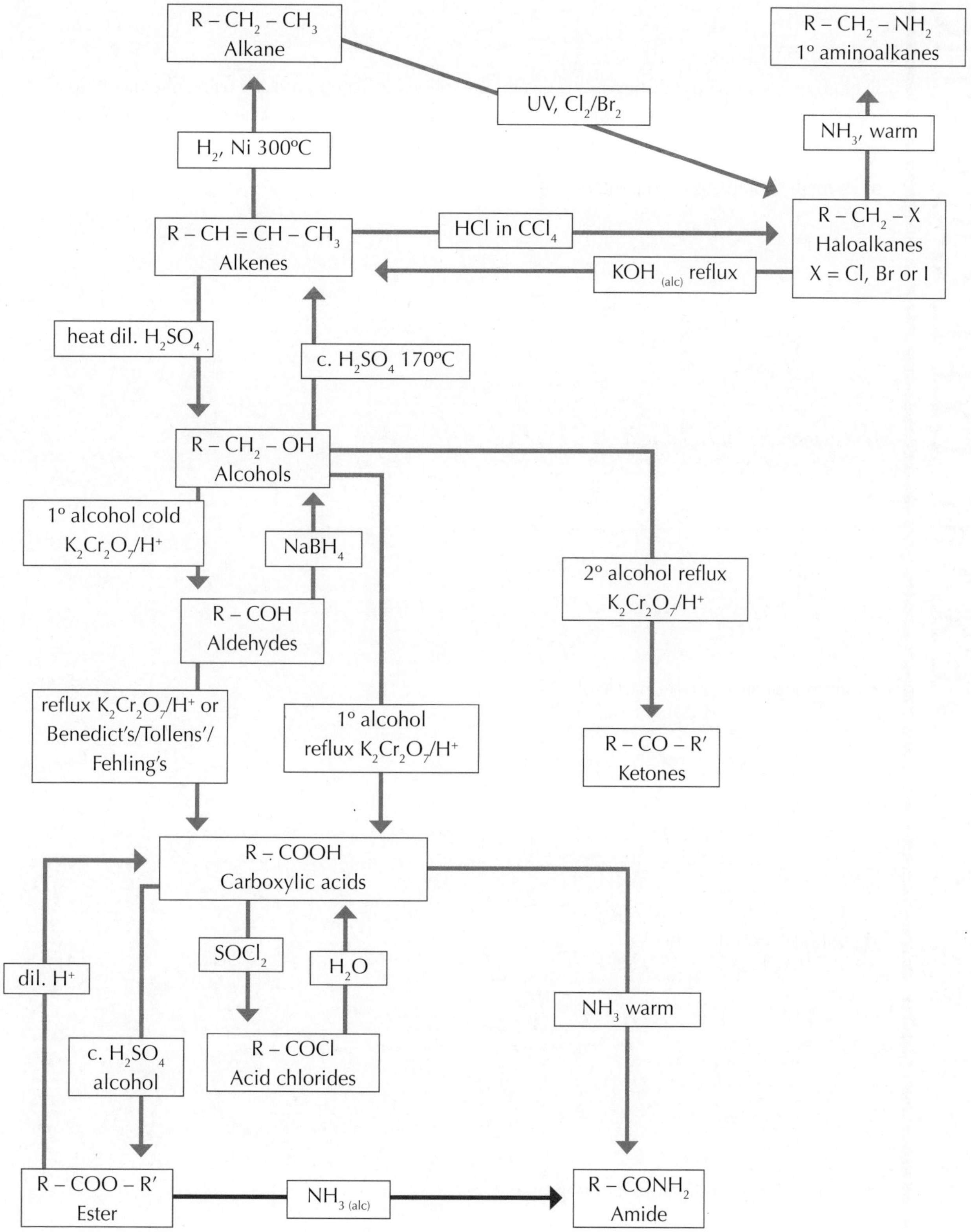

EXAM-TYPE QUESTIONS

Before you get started here are some points to remember from the Examiner's Report:
- Don't use generalisations; use terms carefully.
- Plan your answer before you start.

Question one

Explain how you would distinguish between the following compounds by using chemical means in the laboratory. Give any expected observations, linked to the species involved, and relevant equations. Explain the reasoning behind the tests.

a 2-methylbutan-2-ol and pentan-1-ol

b ethanol and ethanoic acid

c aminopropane and propan-1-ol

d butanal and butanone

e propanoyl chloride and 1-chloropropane

 ISBN: 9780170352611

EXAM-TYPE QUESTIONS

Question two

a Complete the table below by filling in the IUPAC (International Union of Pure and Applied Chemistry) name or drawing the structural formula of the following molecules.

Name	Structure
propanoyl chloride	
2-methylpropanal	
	$CH_3C(O)CH(CH_3)_2$
	$CH_3CH(CH_3)CH(CH_3)CONH_2$
propyl propanoate	

b Explain in terms of the structure and bonding of propyl propanoate why it has a much lower boiling point than hexanoic acid.

c Explain why 2-chloropropanal can exist as optical isomers but 3-chloropropanal cannot. In your answer include 3D drawings of the enantiomers.

Question three

a Complete the table below by filling in the IUPAC name or drawing the structural formula of the following molecules.

Name	Structure
2-methylpentanoic acid	
	$CH_3CH_2CH_2CH(CH_3)COH$
2-bromopentan-3-one	
	$CH_3CH_2C(CH_3)_2CH_2NH_2$
	$CH_3CH_2CH(CH_3)CH_2COCl$

b Describe a physical and a chemical test including any relevant observations and equations that could be used to distinguish between 2-methylpentanoic acid and 2-bromopentan-3-one.

 ISBN: 9780170352611

Question four

a Name and draw the structures of the products of the reactions below with the following molecule:

$$CH_3CH_2C \overset{O}{\underset{NH_2}{\diagdown}}$$

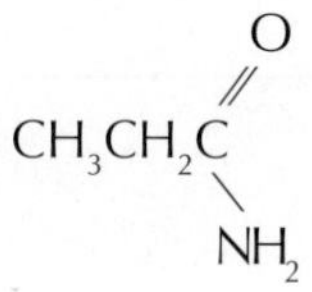

i With heat and hydrochloric acid.

ii With heat and sodium hydroxide.

iii Explain why the two reactions above are considered hydrolysis reactions.

b In the following reaction between methanol and ethanoic acid, if we add concentrated sulfuric acid, what product will we get and what will it be called?

$$H-\underset{\underset{H}{|}}{\overset{\overset{H}{|}}{C}}-O-H \quad + \quad H-\underset{\underset{H}{|}}{\overset{\overset{H}{|}}{C}}-\underset{\underset{O-H}{\diagdown}}{\overset{\diagup\!\diagup O}{C}} \quad \longrightarrow$$

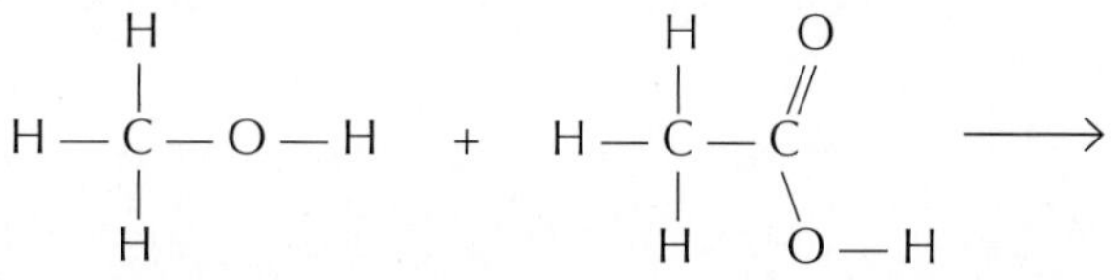

i Name of product: _____

ii Structure of product: _____

iii Explain what kind of reaction the above reaction between methanol and ethanoic acid is.

Question five

a Fill in the blanks in the reaction scheme shown below.

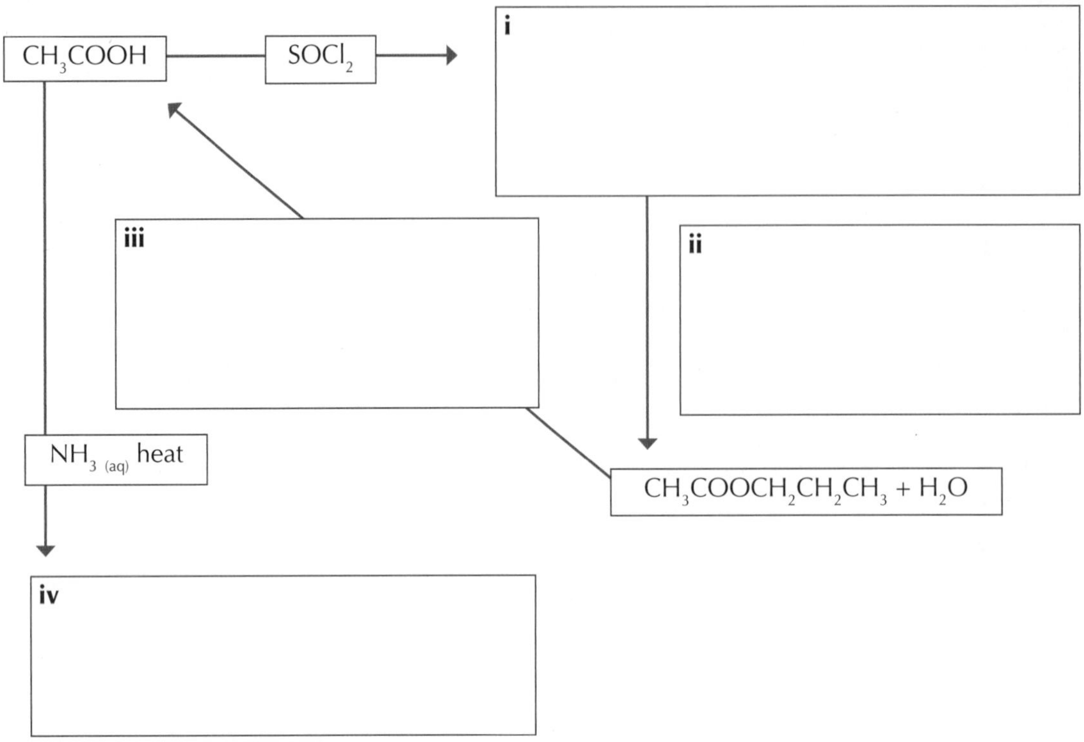

b Explain the different conditions required and the different product formed when the following alcohols react with acidified potassium dichromate, $H^+/K_2Cr_2O_7$.

2-methylpropan-1-ol

2-methylpropan-2-ol

butan-1-ol

ISBN: 9780170352611

EXAM-TYPE QUESTIONS

Question six

Compound **A**, C_4H_8O, reacted with Benedict's solution to produce compound **B** and also did with $NaBH_4$ to produce compound **C**. Compound **C** reacted with $SOCl_2$ to produce compound **D**.

On reacting with alcoholic KOH under reflux, compound **D** produced compound **E**. Compound **E** reacted with HCl in CCl_4 to produce compound **B** and compound **F**, which is optically active. Compound **F** reacts with warm NH_3 to form the optically active compound **G**.

Identify compounds **A** to **G**, giving your reasoning for each decision.

Question seven

Compound **A**, $C_5H_{12}O_2$, reacted with dilute $SOCl_2$ to produce compound **B**, $C_5H_{11}OCl$. Compound **B** then reacted with alcoholic NH_3 and heat to produce compound **C**, $C_5H_{11}ON$. Compound **C** reacted with HCl to produce compound **A** and ammonium chloride. Compound **A** then reacted with compound **D**, C_3H_8O, and concentrated H_2SO_4 to produce compound **E**, $C_8H_{18}O_2$. Compound **D** was heated to 170°C with concentrated H_2SO_4 to produce compound **F**, C_3H_6. Compound **F** was then reacted with HCl in CCl_4 to produce compounds **G** and **H**, where **H** was optically active.

Identify all compounds **A** to **H**, giving your justification for each.

 ISBN: 9780170352611

Question eight

a i Show one of the dipeptides that is formed when the two amino acids below react.

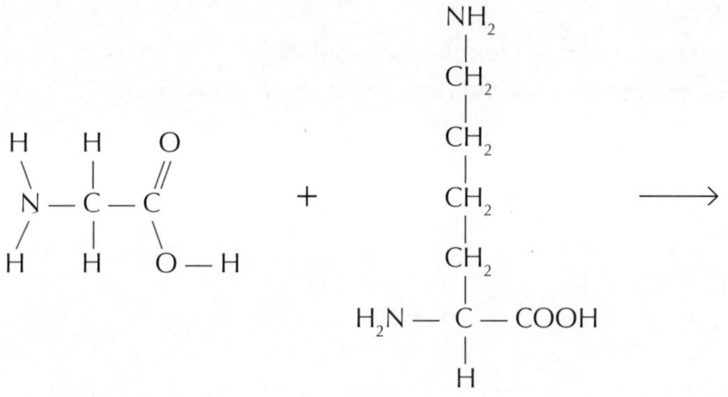

Glycine Lysine

ii Draw a circle around the amide linkage in the dipeptide you have drawn.

iii Explain what type of reaction the above equation shows.

b Draw the amino acid monomers from the dipeptide shown below.

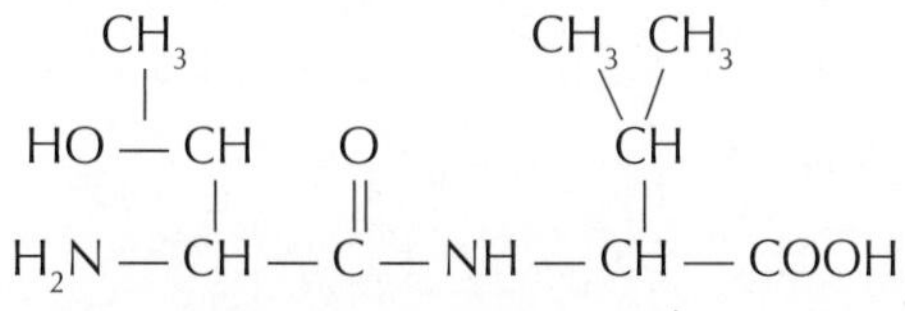

thr-val (threonylvaline)

Question nine

Outline how you would distinguish between the following substances using only litmus paper, water and Fehling's solution. In your answer you should include:

- the name of each molecule
- the test(s) that you would carry out to identify each molecule
- any equations, if applicable, to identify the organic products formed.

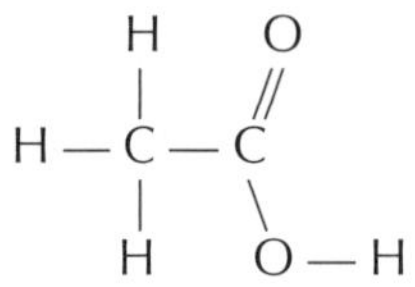
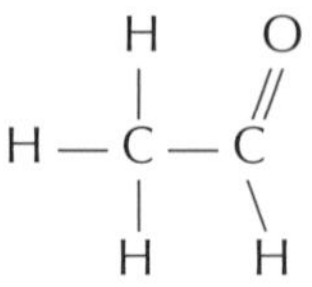

ISBN: 9780170352611

Question ten

Outline how you would distinguish between the following substances using only litmus paper and water. In your answer you should include:

- the name of each molecule
- the test(s) that you would carry out to identify each molecule
- any equations, if applicable, to identify the organic products formed.

$$CH_3CH_2\overset{\displaystyle O}{\underset{\displaystyle NH_2}{C}} \qquad CH_3-\overset{\displaystyle NH_2}{\underset{}{CH}}-CH_3 \qquad H-\overset{\displaystyle H}{\underset{\displaystyle H}{C}}-\overset{\displaystyle H}{\underset{\displaystyle H}{C}}-\overset{\displaystyle O}{\underset{\displaystyle Cl}{C}}$$

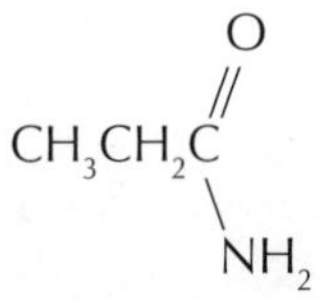

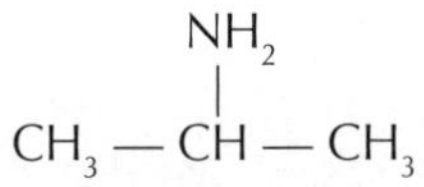

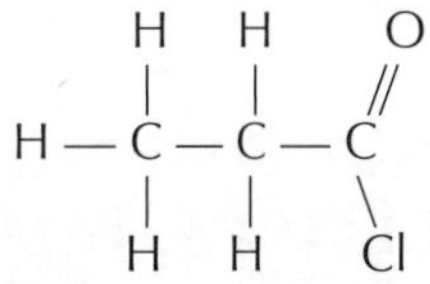

3.6 Demonstrate understanding of equilibrium principles in aqueous systems (91392)

Learning outcomes

Tick off when you have studied these ideas in class and when you have revised that section prior to your assessment.

	Learning outcomes	In class	Revision
1	Relate solubility to K_s.		
2	Calculate the solubility of sparingly soluble ionic solids (limited to AB, A_2B and AB_2 types where neither of the ions A nor B reacts further with water).		
3	Calculate the solubility of solids in water and in solutions already containing one of the ions A or B (a common ion) or due to the formation of a complex ion, or the reaction of a basic anion with added acid.		
4	Predict whether an ionic solution will precipitate or dissolute.		
5	Write qualitative descriptions of relative concentrations of species in a solution.		
6	Calculate the pH and pK_a of a weak acid.		
7	Calculate the pH of a weak base.		
8	Relate the strength of an acid or base to conductivity or pH value.		
9	Describe the properties of a buffer and describe changes made to it by the addition of small amounts of acids or bases.		
10	Draw and interpret a titration curve of either a strong acid with a strong base, a strong acid with a weak base, or a strong base with a weak acid.		

ISBN: 9780170352611

Pre-test — What do you know?

1 Write the equilibrium constant expression, K_c, for the following equilibria.

a $PCl_{3\,(g)} + Cl_{2\,(g)} \rightleftharpoons PCl_{5\,(g)}$

b $2N_2O_{5\,(g)} \rightleftharpoons 4NO_{2\,(g)} + O_{2\,(g)}$

c $SO^{2-}_{3\,(aq)} + H_2O_{\,(l)} \rightleftharpoons HSO^{-}_{3\,(aq)} + OH^{-}_{\,(aq)}$

d $CO_{\,(g)} + H_2O_{\,(g)} \rightleftharpoons CO_{2\,(g)} + H_{2\,(g)}$

2 Explain what effect (if any) the following changes would have on the equilibria below using Le Chatelier's principle and any effect to the K_c value to explain your answer.

$4NH_{3\,(g)} + 5O_{2\,(g)} \rightleftharpoons 4NO_{\,(g)} + 6H_2O_{\,(g)}$ $\quad \Delta_r H^{o} = -46.2\ kJmol^{-1}$

a Increasing the temperature.

b Increasing the total pressure.

c Adding a catalyst.

d Decreasing the ammonia, NH_3, concentration.

3 Complete the following reactions of acids reacting with water.

a $HCl + H_2O \longrightarrow$

b $H_2SO_4 + H_2O \longrightarrow$

c $CH_3CH_2COOH + H_2O \rightleftharpoons$ ______

d $Mg(OH)_{2\ (aq)} \longrightarrow$ ______

e $NH_3 + H_2O \rightleftharpoons$ ______

4 Calculate the pH of the following solutions of acid.

a 1 $molL^{-1}$ hydrochloric acid solution ______

b 0.0122 $molL^{-1}$ hydrochloric acid solution ______

5 Calculate the pH of the following basic solutions.

a 0.100 $molL^{-1}$ sodium hydroxide solution ______

b 1.20 $molL^{-1}$ sodium hydroxide ______

6 Explain the difference between the strength and conductivity of 0.1 $molL^{-1}$ solutions of propanoic acid and sulfuric acid.

ISBN: 9780170352611

Recapping Level 2

In Level 2 you studied equilibria and the properties of acids and bases. The aqueous standard covers these ideas in more detail and extends them to new levels. So before we get started on this chapter, let's review the work covered last year.

Equilibria

An equilibrium is a reaction which can be reversed. We can make changes to a dynamic equilibrium (the point where the rate of reaction of the reactants going to the products equals the rate of reaction of the products going into the reactants) by changing the concentration, temperature, pressure or by adding a catalyst. Any change made to an equilibrium system will force the equilibrium to alter in order to correct the change made to the system (Le Chatelier's principle). Remember, though, only temperature will alter the value of K, the equilibrium constant.

Example:

$$N_{2\,(g)} + 3H_{2\,(g)} \longrightarrow 2NH_{3\,(g)}$$

The equilibrium above will shift towards the products if pressure is increased as the system corrects itself from the changes made to it. This does not alter the value of K, the equilibrium constant, as there will still be the same ratio of products to reactants.

If we increase the temperature, though, K will decrease in value as the forward reaction is exothermic and so the equilibrium will shift towards the reactants in order to correct the change made to the system, by lowering the temperature of the surroundings by shifting in the endothermic direction of the reaction.

Remember, $K = \frac{[\text{products}]}{[\text{reactants}]}$ where [ ] = concentration of in $molL^{-1}$

Acids and bases

An acid is a proton donor and it has a pH value less than 7. We can calculate the pH of a strong acid using the formula pH = $-\log[H_3O^+]$; the concentration of a strong acid will equal the concentration of hydronium ions, H_3O^+, since they fully dissociate in water.

Example:

$$HCl + H_2O \longrightarrow Cl^- + H_3O^+$$

If the concentration of HCl is 0.1 $molL^{-1}$, then the pH = -log 0.1 = 1.

A weak acid like ethanoic acid, CH_3COOH, will only partially dissociate in water and so will have an equilibrium arrow in its reaction with water:

$$CH_3COOH + H_2O \rightleftharpoons CH_3COO^- + H_3O^+$$

ISBN: 9780170352611

A base is a proton acceptor and it has a pH above 7. We can calculate the pH of a strong base using the same formula as above and the water dissociation constant, $K_w = [H_3O^+][OH^-] = 1 \times 10^{-14}$. The concentration of hydroxide ions produced will always equal the concentration of strong base used if it is a strong base, since it will fully dissociate in water.
Example:

$$NaOH_{(aq)} \longrightarrow Na^+ + OH^-$$

If the concentration of NaOH is 0.1 molL^{-1}, $[H_3O^+] = K_w/[OH^-] = 1 \times 10^{-14}/0.1 = 1 \times 10^{-13}$.
$pH = -\log[H_3O^+] = -\log(1 \times 10^{-13}) = 13$

A weak base like ammonia, NH_3 will only partially dissociate in water to produce hydroxide ions, OH^-, and so it will also have an equilibrium arrow to represent its equation with water:

$$NH_3 + H_2O \rightleftharpoons NH_4^+ + OH^-$$

What is aqueous chemistry?

Aqueous chemistry is the study of the solubility of different ionic compounds in water and acids and bases.

The properties of an aqueous system can be altered by changing the concentration and nature of species present.

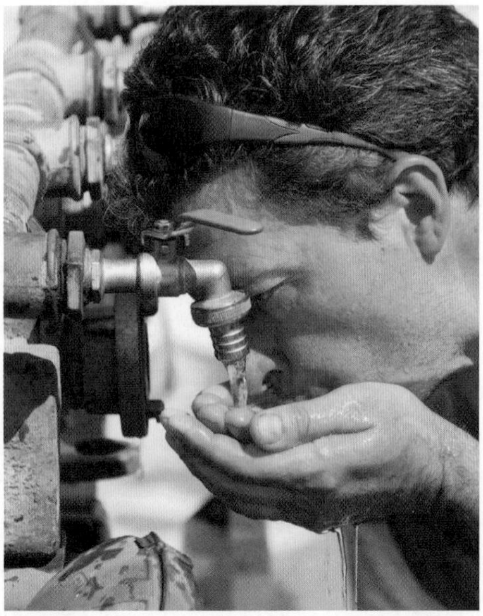

CHEMISTRY APPS

The freezing point of sea water is lower than that for pure water. Water is also less dense in solid state, a fact which is essential to life as it means lakes, rivers, streams and oceans freeze from the top allowing life to continue living below.

EXAMINER'S TIP

Before embarking on the first part of this standard, in order to achieve you must know your formulas for common ions and write them correctly. So if you don't know them now, ask your teacher for a list of common ions and go home tonight and learn them.

ISBN: 9780170352611

EXPERIMENT 1

Solubility

AIM: To look at a solubility equilibrium and see what affects it.

EQUIPMENT and CHEMICALS:

1 boiling tube
test tube holders
lead nitrate, $Pb(NO_3)_2$ (0.1 $molL^{-1}$)
Bunsen burner
heatproof mat
potassium iodide, KI (0.1 $molL^{-1}$)

SAFETY PRECAUTIONS:

Both lead nitrate and potassium iodide are skin and eye irritants, and lead nitrate is also very hazardous if swallowed.

METHOD and RESULTS:

1 Add about 2 mL of potassium iodide to about 2 mL of lead nitrate to a boiling tube.

What happened? ____

____

____

____

2 Gently heat your boiling tube over a blue Bunsen flame by waving the boiling tube in and out of the flame and pointing the mouth of the boiling tube towards the wall.

What happened? ____

____

____

____

CONCLUSION:

What happens when you heat an insoluble substance such as lead iodide? What does it tell you about the type of reaction occurring?

____

____

____

____

____

____

____

____

____

____

____

Solubility equilibria

Solubility equilibria are a specific type of equilibria where an ionic solid dissolves in water to produce its ions. Ionic solids have varying amounts of solubility in water depending on the ions present. Solubility equilibria can be represented by an equilibrium expression, K_s, where the concentrations at equilibrium are used.

Example:

$CaSO_{4\,(s)} \rightleftharpoons Ca^{2+}_{(aq)} + SO_{4\,(aq)}^{2-}$ $K_s (CaSO_4) = 2.4 \times 10^{-5} = [Ca^{2+}][SO_4^{2-}]$ at room temperature and pressure

Remember, the equilibrium expression is always the concentration of the products divided by the concentration of reactants. In solubility equilibria, since the reactant is always a solid, we don't include it in the K_s expression. K_s is the point in the reaction where the rate of the forward reaction is equal to the rate of the reverse reaction and can only be altered by changes in temperature.

The solubility of a salt can be classified as being soluble, sparingly soluble or insoluble. All the solids involved in the following equilibria are sparingly soluble, which means they will partially dissolve in water. Depending on the size of their K_s value, they will dissolve more or less than each other.

3.1: Solubility equilibria

Fill in the table below of solubility equilibria; the first one has been done for you.

Compound	Formula	Equation	K_s expression
Silver iodide	AgI	$AgI_{(s)} \rightleftharpoons Ag^+_{(aq)} + I^-_{(aq)}$	$K_s = [Ag^+][I^-]$
Zinc sulfide			
Copper (II) chloride			
Calcium fluoride			
Lead iodide			
Iron (II) sulfate			
Silver carbonate			
Lead chloride			

ISBN: 9780170352611

Solubility, s

If we know the solubility, s, for a compound and its formula, we can calculate the solubility product or constant.

Example:

The solubility of barium sulfate, $BaSO_4$, is 1.05×10^{-5} $molL^{-1}$. Calculate the solubility product.

The equilibrium is:

$$BaSO_{4\,(s)} \rightleftharpoons Ba^{2+}_{(aq)} + SO^{2-}_{4\,(aq)}$$

When 1 mole of barium sulfate dissolves in water, 1 mole of barium and sulfate ions will be in solution. So that means:

$$[Ba^{2+}] = [SO_4^{2-}] = 1.05 \times 10^{-5} \text{ molL}^{-1}$$

In order to calculate the solubility product, K_s, we place these values into the solubility expression:

$$K_s = [Ba^{2+}][SO_4^{2-}] = 1.05 \times 10^{-5} \times 1.05 \times 10^{-5} = 1.10 \times 10^{-10}$$

Note: you don't need to write the units for the K_s expression, but s has the unit $molL^{-1}$ and you must include it.

Types of ionic compounds

Barium sulfate, $BaSO_4$, is also known as an AB compound as the ions are one-to-one in the chemical formula. Whenever we have an AB compound, the solubility product can always be calculated by the following algebraic expression:

$$K_s(AB) = [A^+][B^-] = s \times s = s^2$$

We can rearrange this formula in order to calculate the solubility, or s:

$$s = \sqrt{K_s}$$

There are also A_2B compounds, for example silver sulfide, Ag_2S:

$$K_s(A_2B) = [A^+]^2[B^{2-}] = (2s)^2 \times s = 4s^3$$

And AB_2 compounds, for example magnesium hydroxide, $Mg(OH)_2$:

$$K_s(AB_2) = [A^{2+}][B^-]^2 = s \times (2s)^2 = 4s^3$$

Rearranged this is:

$$s = \sqrt[3]{K_s/4}$$

ISBN: 9780170352611

EXAMINER'S TIP

You must show all your working for these calculations and not round your answer until the final step, as correct working can still earn you marks. No working and an incorrect answer will give you nothing.

3.2: Solubility product calculations

Calculate the solubility products for the following equilibria; the first one has been done for you.

1 Calculate the solubility product of magnesium hydroxide, $Mg(OH)_2$, given the solubility is 1.71×10^{-4} $molL^{-1}$. In your answer include:

- the equilibrium equation
- the equilibrium expression
- the calculation of the solubility product.

$$Mg(OH)_{2\,(s)} \rightleftharpoons Mg^{2+}_{(aq)} + 2OH^-_{(aq)}$$

$$K_s(Mg(OH)_2) = [Mg^{2+}][OH^-]^2 = 4s^3$$

$$K_s(Mg(OH)_2) = 4 \times (1.71 \times 10^{-4})^3 = 2.00 \times 10^{-11}$$

2 Calculate the solubility product of silver chloride, AgCl, given the solubility is 1.60×10^{-3} $molL^{-1}$. In your answer include:

- the equilibrium equation
- the equilibrium expression
- the calculation of the solubility product.

3 Calculate the solubility product of lead iodide, PbI_2, given the solubility is 1.51×10^{-4} $molL^{-1}$. In your answer include:

- the equilibrium equation
- the equilibrium expression
- the calculation of the solubility product.

ISBN: 9780170352611

4 Calculate the solubility product of silver sulfate, Ag_2SO_4, given the solubility is 1.31×10^{-5} $molL^{-1}$. In your answer include:

- the equilibrium equation
- the equilibrium expression
- the calculation of the solubility product.

5 Calculate the solubility product of copper carbonate, $CuCO_3$, given the solubility is 1.2×10^{-5} $molL^{-1}$. In your answer include:

- the equilibrium equation
- the equilibrium expression
- the calculation of the solubility product.

6 Calculate the solubility of barium sulfate, $BaSO_4$, given the solubility product $K_s(BaSO_4) = 1.1 \times 10^{-10}$. In your answer include:

- the equilibrium equation
- the equilibrium expression
- the calculation of the solubility.

7 Calculate the solubility of silver iodide, AgI, given the solubility product $K_s(AgI) = 8.3 \times 10^{-17}$. In your answer include:

- the equilibrium equation
- the equilibrium expression
- the calculation of the solubility.

ISBN: 9780170352611

8 Calculate the solubility of copper hydroxide, $Cu(OH)_2$, given the solubility product $K_s(Cu(OH)_2) = 4.8 \times 10^{-20}$. In your answer include:

- the equilibrium equation
- the equilibrium expression
- the calculation of the solubility.

9 Calculate the solubility of calcium fluoride, CaF_2, given the solubility product $K_s(CaF_2) = 3.45 \times 10^{-11}$. In your answer include:

- the equilibrium equation
- the equilibrium expression
- the calculation of the solubility.

10 Calculate the solubility of magnesium fluoride, MgF_2, given the solubility product $K_s(MgF_2) = 5.16 \times 10^{-11}$. In your answer include:

- the equilibrium equation
- the equilibrium expression
- the calculation of the solubility.

CHEMISTRY APPS

Why is solubility of ionic compounds important?

We use the solubility of substances in order to separate out one from the other and it is one of the key properties when we describe the nature of a solid.

It is also important when looking at how a pollutant will affect an environment or how a drug will work within the body.

 ISBN: 9780170352611

The common ion effect

A sparingly soluble solid will be less soluble if there is the presence of a common ion (an ion that is present already in the equilibrium).

For example, if you have a saturated solution of AgCl and you add HCl to it, you have a new $[Cl^-]$. Since K_s must remain constant, the equilibrium will shift to the left. Therefore less AgCl will dissolve.

Example:

What is the solubility of $CaCO_3$ when you add 0.100 $molL^{-1}$ of K_2CO_3?

$CaCO_{3\,(s)} \rightleftharpoons Ca^{2+}_{(aq)} + CO^{2-}_{3\,(aq)}$

$K_s = 3.3 \times 10^{-9}$

If the solubility of the $CaCO_3$ is then:

$[Ca^{2+}] = s$ $molL^{-1}$

$[CO_3^{2-}] = (0.100 + s)$ $molL^{-1}$

Assumption that s is much less than 0.100 $molL^{-1}$.

$s = 3.3 \times 10^{-9}/0.100 = 3.3 \times 10^{-8}$ $molL^{-1}$

Note you must always write in the assumption as part of your working for this type of question.

3.3: The common ion effect

1 Calculate the solubility of barium carbonate, $BaCO_3$, given the following information.

$K_s(BaCO_3) = 5 \times 10^{-9}$ $[K_2CO_3] = 0.300$ $molL^{-1}$

In your answer you should include:

- the equilibrium reaction
- the K_s expression
- the assumption you make
- the calculation and all working for calculating the solubility.

2 Calculate the solubility of lead sulfate, $PbSO_4$, given the following information.

$K_s(PbSO_4) = 2 \times 10^{-8}$ $[Na_2SO_4] = 0.0501\ molL^{-1}$

In your answer you should include:
- the equilibrium reaction
- the K_s expression
- the assumption you make
- the calculation and all working for calculating the solubility.

3 Calculate the solubility of calcium fluoride, CaF_2, given the following information.

$K_s(CaF_2) = 4 \times 10^{-11}$ $[Ca(NO_3)_2] = 0.100\ molL^{-1}$

In your answer you should include:
- the equilibrium reaction
- the K_s expression
- the assumption you make
- the calculation and all working for calculating the solubility.

ISBN: 9780170352611

4 Calculate the solubility of lead iodide, PbI_2, given the following information.

$$K_s(PbI_2) = 1 \times 10^{-9} \qquad [KI] = 0.0101 \text{ molL}^{-1}$$

In your answer you should include:

- the equilibrium reaction
- the K_s expression
- the assumption you make
- the calculation and all working for calculating the solubility.

5 Calculate the solubility of silver iodide, AgI, given the following information.

$$K_s(AgI) = 8 \times 10^{-17} \qquad [AgNO_3] = 1.10 \times 10^{-3} \text{ molL}^{-1}$$

In your answer you should include:

- the equilibrium reaction
- the K_s expression
- the assumption you make
- the calculation and all working for calculating the solubility.

What do the following key words mean?

Solubility	
Insoluble	
Sparingly soluble	
Soluble	
Solubility product	
Ionic product	

CHECKPOINT 1
WHAT HAVE YOU LEARNED SO FAR?

Solubility and the common ion effect

1 Calculate the solubility of the following compounds.

a $K_s(AlPO_4) = 9.84 \times 10^{-21}$

b $K_s(BaF_2) = 1.84 \times 10^{-7}$

c $K_s(PbBr_2) = 6.60 \times 10^{-6}$

 ISBN: 9780170352611

CHECKPOINT 1
WHAT HAVE YOU LEARNED SO FAR?

d $K_s(KIO_4) = 3.71 \times 10^{-4}$

e $K_s(Sn(OH)_2) = 5.45 \times 10^{-27}$

2 Calculate the solubility product for the following solubility values.

a For $ZnCO_3$ where $s = 1.21 \times 10^{-5}$ $molL^{-1}$

b For $Ni(OH)_2$ where $s = 5.16 \times 10^{-6}$ $molL^{-1}$

c For HgI_2 where $s = 1.94 \times 10^{-10}$ $molL^{-1}$

d For Ag_2CO_3 where $s = 1.28 \times 10^{-4}$ $molL^{-1}$

e For $SrSO_4$ where $s = 5.87 \times 10^{-4}$ $molL^{-1}$

ISBN: 9780170352611

3 Calculate the new solubility, s, when the following equilibria have common ions added to them.

a $K_s(HgS) = 2 \times 10^{-53}$ $c(Na_2S) = 0.105$ $molL^{-1}$

b $K_s(MnCO_3) = 2.24 \times 10^{-11}$ $c(Na_2CO_3) = 0.166$ $molL^{-1}$

c $K_s(FeF_2) = 2.36 \times 10^{-6}$ $c(Fe(NO_3)_2) = 0.201$ $molL^{-1}$

d $K_s(Cu(OH)_2) = 4.8 \times 10^{-20}$ $c(KOH) = 1.22$ $molL^{-1}$

e $K_s(CdF_2) = 6.44 \times 10^{-3}$ $c(KF) = 0.0607$ $molL^{-1}$

 ISBN: 9780170352611

Predicting precipitation

If we mix two solutions together with a common ion present we can predict whether a precipitate will form from the equilibrium shift.

We can do this using Q_s, the ionic product; this is the reaction quotient when an equilibrium has not yet been established. The expression for Q_s is the same as for K_s, the solubility product.

If, after mixing two solutions together, Q_s is greater than K_s, there is an excess of products and precipitation will occur.

If Q_s equals K_s, there is a saturated solution (that means no more solid will dissolve in solution).

If Q_s is less than K_s, no precipitation occurs.

Example:

If equal volumes of 0.100molL^{-1} barium chloride and 0.0500 molL^{-1} sodium sulfate are added, predict whether any barium sulfate, $BaSO_4$, will be precipitated.

$$K_s(BaSO_4) = 1.08 \times 10^{-10}$$

On mixing, the concentration halves as there is double the volume:

$$[Ba^{2+}] = 0.100/2 = 0.0500 \text{ molL}^{-1}$$
$$[SO_4^{2-}] = 0.0500/2 = 0.0250 \text{ molL}^{-1}$$

Using the K_s expression to calculate Q_s:

$$Q_s = [Ba^{2+}][SO_4^{2-}] = 0.0500 \times 0.0250 = 0.00125$$

Since $Q_s > K_s$, precipitation will occur.

3.4: Predicting precipitation

Predict whether a precipitate will form in the following reactions. In your answer, include the following:

- an equilibrium reaction
- a Q_s expression with calculation
- a statement as to whether Q_s is larger, equal to or smaller than K_s.

a 55 mL of 5.61 x 10^{-3} molL^{-1} copper (II) nitrate, $Cu(NO_3)_2$, mixed with 155 mL of 6.11 x 10^{-5} molL^{-1} sodium sulfide, Na_2S, solution. $K_s(CuS) = 8 \times 10^{-37}$

ISBN: 9780170352611

b 65 mL of 0.0501 $molL^{-1}$ iron nitrate, $Fe(NO_3)_2$, mixed with 45 mL of 0.00102 $molL^{-1}$ sodium fluoride, NaF, solution. $K_s(FeF_2) = 2.36 \times 10^{-6}$

c 45 mL of 0.000601 $molL^{-1}$ magnesium nitrate, $Mg(NO_3)_2$, mixed with 45 mL of 0.000702 $molL^{-1}$ sodium carbonate, Na_2CO_3, solution. $K_s(MgCO_3) = 6.82 \times 10^{-6}$

d 75 mL of 0.00101 $molL^{-1}$ silver nitrate, $AgNO_3$, mixed with 145 mL of 0.00602 $molL^{-1}$ sodium chloride, NaCl, solution. $K_s(AgCl) = 1.77 \times 10^{-10}$

e 65 mL of 0.100 $molL^{-1}$ zinc nitrate, $Zn(NO_3)_2$, mixed with 155 mL of 0.0202 $molL^{-1}$ sodium hydroxide, NaOH, solution. $K_s(Zn(OH)_2) = 3 \times 10^{-17}$

ISBN: 9780170352611

f 100 mL of 0.000502 $molL^{-1}$ cobalt nitrate, $Co(NO_3)_2$, mixed with 57 mL of 0.000141 $molL^{-1}$ sodium carbonate, Na_2CO_3, solution. $K_s(CoCO_3) = 1.0 \times 10^{-10}$

g 165 mL of 0.00301 $molL^{-1}$ lead nitrate, $Pb(NO_3)_2$, mixed with 175 mL of 0.00607 $molL^{-1}$ sodium iodide, NaI, solution. $K_s(PbI_2) = 9.8 \times 10^{-9}$

Complex ions and solubility

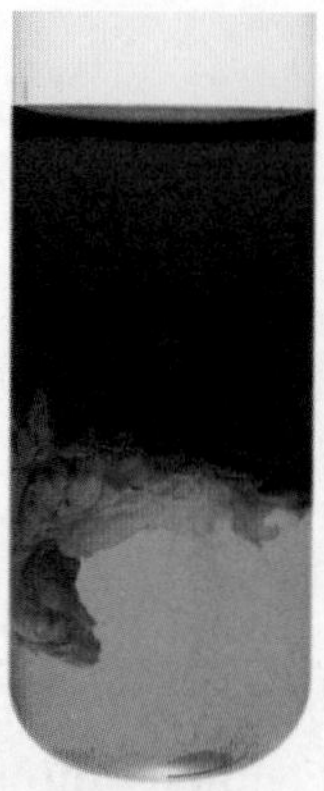

Sometimes ions can be removed from an equilibrium due to the formation of a complex ion. A complex ion is an ion that forms with a metal ion and ligands (such as water or ammonia or they could be an atom that is bound to the metal ion), making an ion with multiple atoms involved.

You may have encountered these last year when you completed the internal on Precipitations.

Common ones you should be able to recognise are shown below.

Complex ion	When it occurs
$[Ag(NH_3)_2]^+$	When ammonia, NH_3, is added in excess to silver chloride, AgCl, or silver oxide, Ag_2O
$[Cu(NH_3)_4]^{2+}$ — this one has a deep-blue colour	When ammonia, NH_3, is added in excess to copper hydroxide, $Cu(OH)_2$
$[Zn(OH)_4]^{2-}$	When excess sodium hydroxide, NaOH, is added to zinc hydroxide, $Zn(OH)_2$
$[Zn(NH_3)_4]^{2+}$	When excess ammonia, NH_3, is added to zinc hydroxide, $Zn(OH)_2$
$[Pb(OH)_4]^{2-}$	When excess sodium hydroxide, NaOH, is added to lead hydroxide, $Pb(OH)_2$

ISBN: 9780170352611

EXPERIMENT 2

Complex ions

AIM: To observe the effects of the formation of a complex ion on a solubility equilibrium.

EQUIPMENT and CHEMICALS:

5 test tubes
test tube rack
0.1 $molL^{-1}$ silver nitrate
1 $molL^{-1}$ sodium hydroxide
1 $molL^{-1}$ ammonia
0.1 $molL^{-1}$ copper nitrate
0.1 $molL^{-1}$ zinc nitrate
0.1 $molL^{-1}$ lead nitrate
0.1 $molL^{-1}$ sodium chloride
plastic dropper/pipette

SAFETY PRECAUTIONS:

Sodium hydroxide and ammonia are corrosive. Silver nitrate is an irritant to the skin and if inhaled, and can cause blindness if it enters the eye. Copper nitrate, zinc nitrate and lead nitrate are all skin and eye irritants and harmful if swallowed. Sodium chloride is a skin and eye irritant.

METHOD and RESULTS:

1 In order to form the silver complex ion, add 1 mL of silver nitrate to a test tube and a few drops of sodium chloride. Record what happens in the table below, then add excess ammonia and record what happens.

$AgNO_3$ + NaCl; observations and name of precipitate formed	AgCl + NH_3; observations and formula of precipitate formed

2 In order to form the copper complex ion, add 1 mL of copper nitrate to a test tube and a few drops of ammonia. Record what happens in the table below, then add excess ammonia and record what happens.

$Cu(NO_3)_2$ + NH_3; observations and name of precipitate formed	$Cu(OH)_2$ + excess NH_3; observations and formula of precipitate formed

ISBN: 9780170352611

3 In order to form the zinc complex ion, add 1 mL of zinc nitrate to a test tube and a few drops of sodium hydroxide. Record what happens in the table below, then add excess sodium hydroxide and record what happens.

$Zn(NO_3)_2$ + NaOH; observations and name of precipitate formed	$Zn(OH)_2$ + excess NaOH; observations and formula of precipitate formed

4 In order to form the other zinc complex ion, add 1 mL of zinc nitrate to a test tube and a few drops of ammonia. Record what happens in the table below, then add excess ammonia and record what happens.

$Zn(NO_3)_2$ + NH_3; observations and name of precipitate formed	$Zn(OH)_2$ + excess NH_3; observations and formula of precipitate formed

5 In order to form the lead complex ion, add 1 mL of lead nitrate to a test tube and a few drops of sodium hydroxide. Record what happens in the table below, then add excess sodium hydroxide and record what happens.

$Pb(NO_3)_2$ + NaOH; observations and name of precipitate formed	$Pb(OH)_2$ + excess NaOH; observations and formula of precipitate formed

EQUATIONS:

Write equations for every precipitate and complex ion formed.

1

2

3

4

5

3.5: Complex ions and solubility

1 Explain the effect the addition of NH_3 will have on the solubility of AgCl.

2 If $NH_3(aq)$ is added to a solution of Cu^{2+}, a precipitate forms then disappears as more NH_3 is added. Explain these observations.

ISBN: 9780170352611

3 At what pH will 0.1 $molL^{-1}$ Zn^{2+} form a precipitate? Discuss the effect of lowering the pH and of increasing the pH. $K_s(Zn(OH)_2) = 3 \times 10^{-17}$

4 Discuss the effect of adding acid and alkali to a saturated solution of $Pb(OH)_2$. $K_s(Pb(OH)_2) = 6 \times 10^{-16}$

Species in solution

When a soluble substance dissolves in water, the number and concentration of the species formed depends on the nature of the substance.

Remember that H_2O is always present in massive excess (55.6 $molL^{-1}$). Also, water auto-ionises, so OH^- and H_3O^+ will always be present in solution. Their relative concentrations depend on the nature of the substance dissolved, but remember that $[OH^-] \times [H_3O^+]$ always equals 1×10^{-14} $molL^{-1}$ (K_w, the water dissociation constant).

Salts are formed when an acid reacts with a base. The pH of the solution depends on the initial reaction.

Reaction between a strong acid and a strong base

Common strong acids you should be able to recognise are HCl, HBr, HNO_3 and H_2SO_4. Common strong bases you should be able to recognise are KOH and NaOH.

The salt formed is neutral. For example:

$$NaOH_{(aq)} + HCl_{(aq)} \longrightarrow NaCl_{(aq)} + H_2O_{(l)}$$

Reaction between a weak acid and a strong base

Common weak acids you should be able to recognise are HF, CH_3COOH and NH_4^+.

The salt formed contains the conjugate base of the weak acid. Solutions of the salt are slightly basic because the conjugate base reacts with water. For example, when NaOH reacts with CH_3COOH:

$$CH_3COOH_{(aq)} + NaOH_{(aq)} \longrightarrow CH_3COO^-Na^+_{(aq)} + H_2O_{(l)}$$

Aqueous solutions of CH_3COONa are slightly basic because the ethanoate ion (the conjugate base) reacts slightly with water:

$$CH_3COO^-_{(aq)} + H_2O_{(l)} \longrightarrow CH_3COOH_{(aq)} + OH^-_{(aq)}$$

Vinegar is CH_3COOH

This is represented in the graph shown below of relative concentrations of either a weak acid or a strong acid in water.

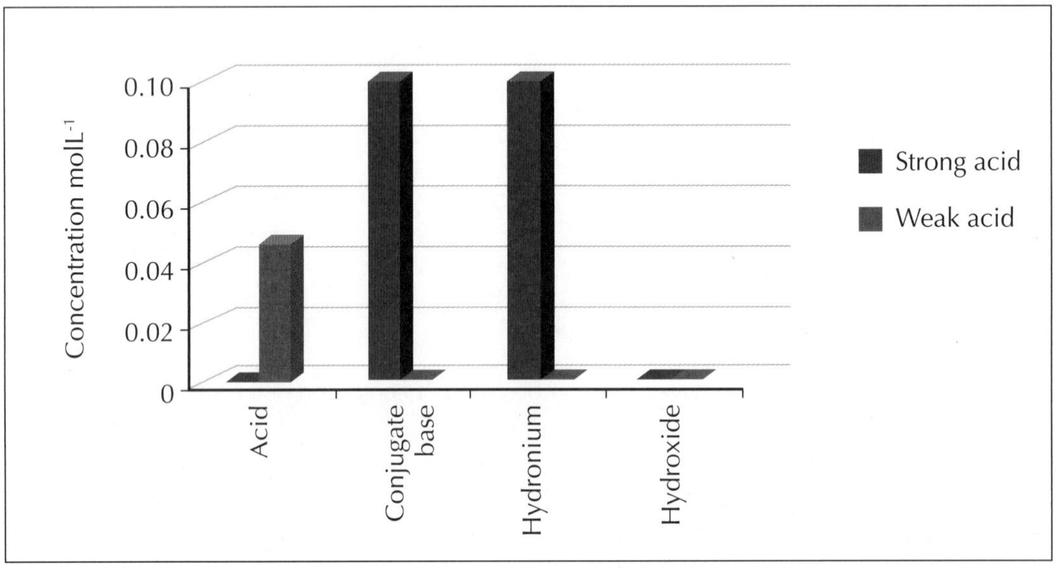

Reaction between a weak base and a strong acid

Common weak bases you should be able to recognise are NH_3, CH_3NH_2 and CH_3COO^-.

The salt formed often contains the conjugate acid of the weak base. Solutions of such salts are slightly acidic because the conjugate acid reacts with water. For example, when HCl reacts with NH_3:

$$HCl_{(aq)} + NH_{3\,(aq)} \longrightarrow NH_4Cl_{(aq)} + H_2O_{(l)}$$

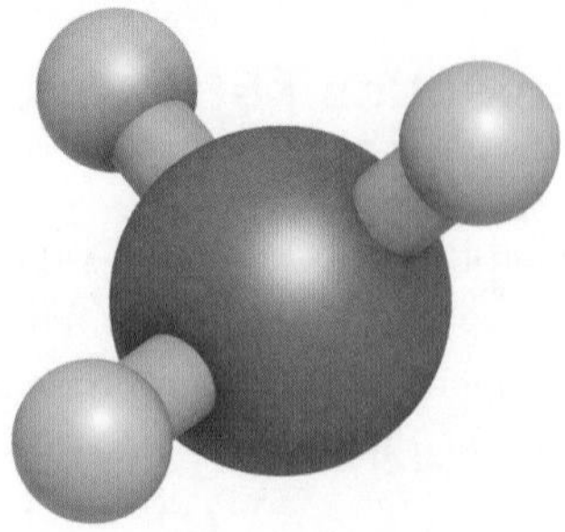

A model of ammonia

Aqueous solutions of NH_4Cl are slightly acidic because the NH_4^+ ion (the conjugate acid) reacts slightly with water:

ISBN: 9780170352611

$$NH_{4\,(aq)}^{+} + H_2O_{(l)} \longrightarrow NH_{3\,(aq)} + H_3O^{+}_{(aq)}$$

Note: When carbonates or hydrogen carbonates react with strong acids, the salt formed does not contain the conjugate acid, so the salt is neutral as shown in the reaction below:

$$2HCl_{(aq)} + Na_2CO_{3\,(aq)} \longrightarrow 2NaCl_{(aq)} + CO_{2\,(g)} + H_2O_{(l)}$$

You will need to be able to state the relative concentrations in order of species in solution.

This is represented in the graph shown below of relative concentrations of either a weak base or a strong base in water.

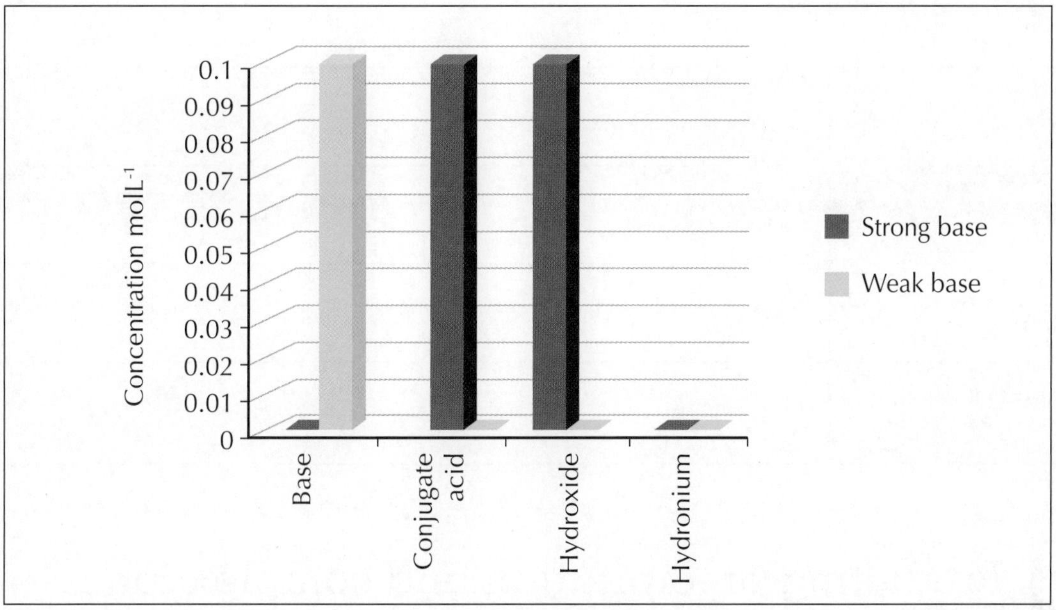

3.6: Species in solution

In the table below, list the order of concentration of species present in solution from highest to lowest; the first one for each type of solution has been done to help you.

Solution	Species present in solution
HCl	$H_3O^+ = Cl^- >>> OH^-$
HNO_3	
HBr	
CH_3COOH	$CH_3COOH > H_3O^+ = CH_3COO^- >> OH^-$
HF	
NaOH	$Na^+ = OH^- >> H_3O^+$
$Mg(OH)_2$	
CH_3NH_2	$CH_3NH_2 > OH^- = CH_3NH_3^+ >> H_3O^+$
CH_3COO^-	

ISBN: 9780170352611

What do the following key words mean?

Precipitate	
Complex ion	
Ligand	
Species	
Strong acid	
Strong base	
Weak acid	
Weak base	
K_w, water dissociation constant	

CHECKPOINT 2

WHAT HAVE YOU LEARNED SO FAR?

Predicting precipitation and complex ions

1 Predict whether a precipitate will occur in the following cases.

a When a solution of 40 mL of 2.01×10^{-4} $molL^{-1}$ calcium nitrate is mixed with a solution of 60 mL of 1.06×10^{-5} $molL^{-1}$ sodium sulfate. $K_s(CaSO_4) = 4.93 \times 10^{-5}$

b When a solution of 55 mL of 1.66×10^{-3} $molL^{-1}$ mercury nitrate is mixed with a solution of 65 mL of 1.06×10^{-3} $molL^{-1}$ sodium iodide. $K_s(HgI_2) = 2.9 \times 10^{-29}$

ISBN: 9780170352611

c When a solution of 115 mL of 6.00×10^{-6} $molL^{-1}$ nickel nitrate is mixed with a solution of 85 mL of 1.11×10^{-3} $molL^{-1}$ sodium carbonate. $K_s(NiCO_3) = 1.42 \times 10^{-7}$

d When a solution of 15 mL of 0.200 $molL^{-1}$ silver nitrate is mixed with a solution of 35 mL of 0.100 $molL^{-1}$ sodium ethanoate. $K_s(AgCH_3COO) = 1.94 \times 10^{-3}$

2 Describe what will happen when the following solutions are added to varying equilibria.

a When ammonia is added to $Cu(OH)_2$ solution.

b When ammonia is added to $Zn(OH)_2$ solution.

c When sodium hydroxide is added to $Zn(OH)_2$ solution.

3 List the species in order of decreasing concentration for the following solutions.

a H_2SO_4

b KOH

c NH_3

d NH_4^+

pH and conductivity of acids

Acids are substances that donate protons.

$$HA_{(aq)} + H_2O_{(l)} \longrightarrow H_3O^+_{(aq)} + A^-_{(aq)}$$

Remember that a strong acid completely dissociates in water, as in the reaction shown above.

A weak acid only partially dissociates in water, that is, it is in equilibrium:

An acid lake in the crater of White Island in the Bay of Plenty.

$$HA_{(aq)} + H_2O_{(l)} \rightleftharpoons H_3O^+_{(aq)} + A^-_{(aq)}$$

As strong acids fully dissociate in water, the concentration of the original acid will equal the concentration of the hydronium ions formed.

For example, if you have 0.1 $molL^{-1}$ of HCl, then there will be 0.1 $molL^{-1}$ Cl^- and H_3O^+.

Then you can work out its pH:

$$pH = -\log_{10}[H_3O^+]$$

This property makes strong acids great conductors of electricity, as there is a high concentration of ions present in their solutions. Weak acids, on the other hand, have lower concentrations of ions and so are generally weaker conductors of current.

EXAMINER'S TIP

You will need to be able to link the pH of an acid to how well it will dissociate in water and therefore how well it will conduct.

This year you will also have to calculate the pH of a weak acid via the following method:

If we have the weak acid HF, then we have the following equilibrium and K_a:

$$HF_{(aq)} + H_2O_{(l)} \rightleftharpoons F^-_{(aq)} + H_3O^+_{(aq)}$$

$$K_a = \frac{[F^-][H_3O^+]}{[HF]}$$ where K_a is the acid dissociation constant

$[F^-] = [H_3O^+]$ so we can rearrange this equation to:

$$K_a = \frac{[H_3O^+]^2}{[HF]}$$

If we assume that very little of HF has dissociated, then we can say the original concentration of HF must be the same, so we can rearrange the above expression to make $[H_3O^+]$ the subject:

$[H_3O^+] = \sqrt{K_a \times [HF]}$ and then calculate the pH using $pH = -\log[H_3O^+]$

ISBN: 9780170352611

CHEMISTRY APPS

The superacid HF/SbF_5 has an acidity higher than 100 per cent pure sulfuric acid and has been shown to protonate the hydrocarbons in a candle, despite the fact that hydrocarbons show no base character.

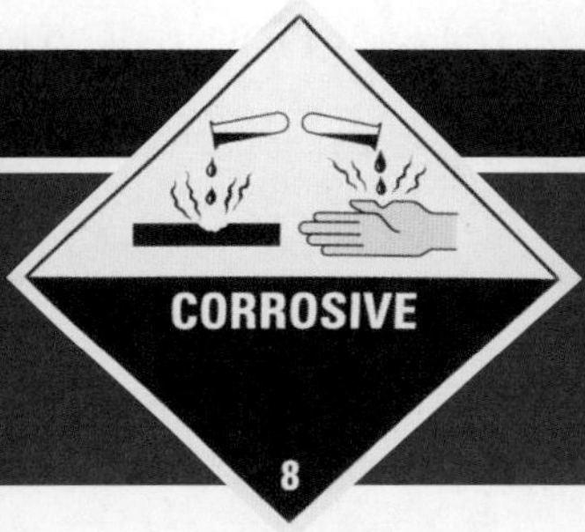

3.7: Calculating the pH of acids

1 Calculate the pH of the following strong acids; include a fully balanced equation in your answer.

a HCl, 0.01 $molL^{-1}$

b HBr, 1.03 $molL^{-1}$

c HNO_3, 0.126 $molL^{-1}$

2 Calculate the pH of the following weak acids; include a K_a expression, fully balanced equation and any assumptions you make in your answer.

a CH_3COOH, 0.0100 $molL^{-1}$, $K_a(CH_3COOH) = 1.8 \times 10^{-5}$

b HF, 0.0100 $molL^{-1}$, $K_a(HF) = 6.3 \times 10^{-4}$

c CH_3CH_2COOH, 0.0100 $molL^{-1}$, $K_a(CH_3CH_2COOH) = 1.3 \times 10^{-5}$

pK_a and K_a

Sometimes you will be given a pK_a instead of a K_a value. A pK_a value is just a $-logK_a$; to make a simpler number, we can convert a pK_a into a K_a using inverse log $-pK_a$.

The smaller the K_a, the weaker the acid, however the higher the pK_a, the weaker the acid.

3.8: pK_a and K_a

1 Convert the following K_a values into pK_a values.

a $K_a(CH_3CH_2COOH) = 1.3 \times 10^{-5}$

b $K_a(HF) = 6.3 \times 10^{-4}$

c $K_a(CH_3COOH) = 1.8 \times 10^{-5}$

2 Convert the following pK_a values into K_a values and then order the acids in terms of increasing strength.

a HNO_2, $pK_a = 3.4$

b HOCl, $pK_a = 7.5$

c HCN, $pK_a = 9.2$

Order of strength:

pH and conductivity of bases

CHEMISTRY APPS

These tufas towers in Mono Lake, USA, are made from limestone ($CaCO_3$), an alkaline substance that forms due to very high salt content found in this lake.

 ISBN: 9780170352611

Just like strong acids, strong bases completely dissociate in water to produce hydroxide ions as they are proton acceptors. This makes them good conductors of current too, as they have lots of ions present in solution.

We can calculate the pH of a strong base using the water dissociation constant, K_w, as shown in the example below.

FACT

An alkali is a soluble base.

> Calculate the pH of 0.0212 $molL^{-1}$ NaOH
>
> $NaOH_{(aq)} \longrightarrow Na^+_{(aq)} + OH^-_{(aq)}$
>
> $K_w = [OH^-][H_3O^+]$
>
> $[H_3O^+] = 1 \times 10^{-14}/0.0212 = 4.71 \times 10^{-13}\ molL^{-1}$
>
> $pH = -\log(4.71 \times 10^{-13}) = 12.3$ (3 s.f.)

When calculating the pH of a weak base, we need to know the K_a for the conjugate weak acid.

Example:

> $NH_{3\,(aq)} + H_2O_{(l)} \rightleftharpoons NH^+_{4\,(aq)} + OH^-_{(aq)}$
>
> NH_4^+ is the conjugate acid of NH_3, as it has an extra proton.

The equation for ammonium reacting with water is:

> $NH^+_{4\,(aq)} + H_2O_{(l)} \rightleftharpoons NH_{3\,(aq)} + H_3O^+_{(aq)}$
>
> $K_a(NH_4^+) = \dfrac{[H_3O^+][NH_3]}{[NH_4^+]} = 5.6 \times 10^{-10}$

We know that $[NH_4^+] = [OH^-]$ and $[OH^-] = K_w/[H_3O^+]$

> So $K_a(NH_4^+) = \dfrac{[NH_3][H_3O^+]}{K_w/[H_3O^+]}$

Rearranging, we get $[H_3O^+] = \sqrt{K_a \times K_w/[NH_3]}$

Finally we can use $pH = -\log[H_3O^+]$

Example:

> Calculate the pH of 0.414 $molL^{-1}$ of CH_3COONa. $K_a(CH_3COOH) = 1.74 \times 10^{-5}$
>
> $CH_3COO^-_{(aq)} + H_2O_{(l)} \rightleftharpoons CH_3COOH_{(aq)} + H_3O^+_{(aq)}$
>
> $K_a(CH_3COOH) = \dfrac{[CH_3COO^-][H_3O^+]}{[CH_3COOH]}$
>
> $[H_3O^+] = \sqrt{((1 \times 10^{-14} \times 1.74 \times 10^{-5})/0.414)} = 6.48 \times 10^{-10}\ molL^{-1}$
>
> $pH = -\log(6.48 \times 10^{-10}) = 9.18$ (3 s.f.)

EXAMINER'S TIP

Students often fail to achieve in this standard because they write incorrect acid-base equations. Practise writing your equations at ***home tonight***.

CHEMISTRY APPS

Acids were first classified by their sourness and bases were classified for their bitterness. Nowadays we don't classify them in this way for obvious reasons (because you shouldn't taste chemicals) but also because some substances can taste sour or bitter and are not acids or bases.

ISBN: 9780170352611

3.9: Calculate the pH of the following bases and explain their conductivity

1 Calculate the pH of the following strong bases and include an equation for its reaction with water in your answer.

a 0.0100 $molL^{-1}$ NaOH

b 0.0105 $molL^{-1}$ KOH

c 1.06×10^{-3} $molL^{-1}$ LiOH

2 Calculate the pH of the following weak bases; include an equation in your answer, the K_a expression and any assumptions you made.

a 0.0100 $molL^{-1}$ F^-, $K_a(HF) = 7.2 \times 10^{-4}$

b 0.0200 $molL^{-1}$ NH_3, $K_a(NH_4^+) = 5.6 \times 10^{-10}$

c 0.00150 $molL^{-1}$ HCO_3^-, $K_a(H_2CO_3) = 4.3 \times 10^{-7}$

ISBN: 9780170352611

d 0.165 $molL^{-1}$ $HCOO^-$, $K_a(HCOOH) = 1.77 \times 10^{-4}$

3 Explain why a 0.100 $molL^{-1}$ solution of NaOH is a better conductor than a 0.100 $molL^{-1}$ solution of CH_3NH_2.

Buffers

Buffer solutions maintain a reasonably constant pH on the addition of small amounts of H_3O^+ and OH^-. They are really important in living organisms, and medical and industrial processes.

Buffers are made from similar concentrations of weak acid and its conjugate base. So upon addition of a base, the acid neutralises the added base, and upon addition of an acid, the base neutralises the added acid.

CHEMISTRY APPS

Blood is a buffer that keeps the pH at 7.4. Any change to the blood's pH that is beyond 0.2 pH units can be lethal, so it is crucially important that its pH is maintained.

Blood is kept at that pH as it is the best value to ensure haemoglobin (the oxygen transport molecule in your blood) functions correctly. Haemoglobin has many equilibria, however overall the following equilibrium is the result:

$$HbH^+_{(aq)} + O_{2\,(aq)} \rightleftharpoons HbO_{2\,(aq)} + H^+_{(aq)}$$

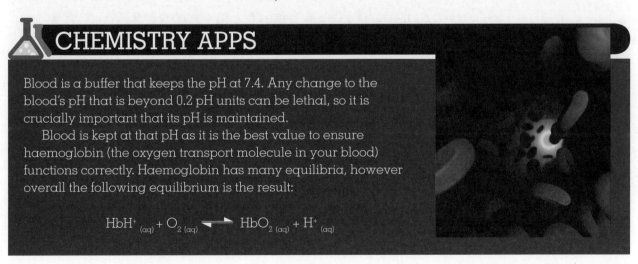

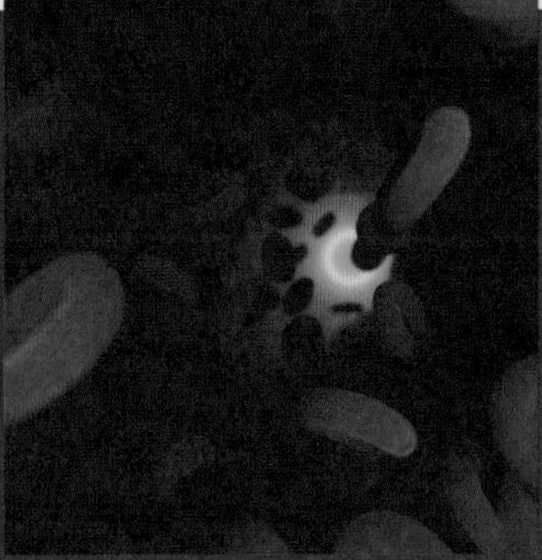

Buffer solutions are effective over a pH range of one pH unit above or below the pK_a value for the weak acid of the buffer. The pH of a buffer solution is unchanged if the solution is diluted or upon the addition of small amounts of acid or base (because the ratio of acid and conjugate base remains the same), however the capacity of the buffer solution will be reduced.

ISBN: 9780170352611

Example:

25 mL of a methanoic acid/sodium methanoate buffer with pH = 3.75 is diluted to 250 mL. pK_a = 3.75.

Original buffer: pH = pKa and [HCOOH] = [NaHCOO]

Diluted buffer: [HCOOH] = x/10 $molL^{-1}$ = [NaHCOO]

pH is unchanged, but the buffer capacity is reduced by a factor of 10.

Calculating the pH of a buffer solution

If a buffer solution contains the same amounts of a weak acid and its conjugate base, then the pH of the buffer is equal to the pK_a of the acid. However, if the concentrations vary, then we need to be able to carry out a calculation.

Example:

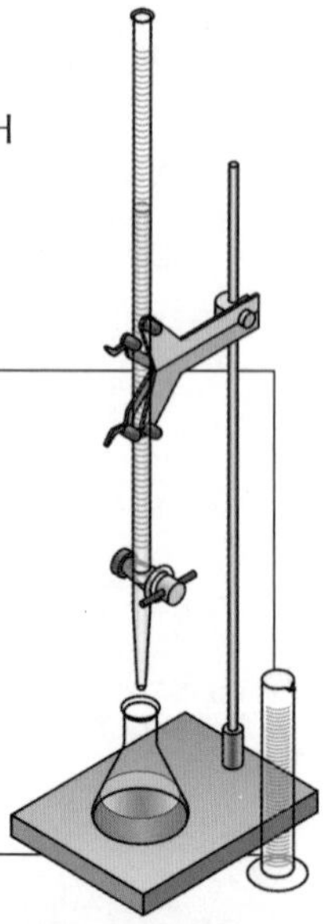

Calculate the pH of a buffer solution when 0.150 $molL^{-1}$ of sodium methanoate, NaHCOO is added to 0.120 $molL^{-1}$ of methanoic acid, HCOOH. $K_a = 1.82 \times 10^{-4}$.

$HCOOH + H_2O \rightleftharpoons HCOO^- + H_3O^+$

$K_a = [HCOO^-][H_3O^+]/[HCOOH]$

$[H_3O^+] = K_a \times [HCOOH]/[HCOO^-]$

$[H_3O^+] = 1.82 \times 10^{-4} \times 0.12/0.15 = 1.456 \times 10^{-4}\ molL^{-1}$

$pH = -\log[H_3O^+] = 3.84$ (3 s.f.)

EXAMINER'S TIP

You will need to write equations to show a buffer reacting with an acid or base. You will also need to be able to relate the pH of the buffer to how much acid or base make it up.

For example, if small amounts of NaOH are added to the methanoic acid/sodium methanoate buffer:

$$NaOH + HCOOH \longrightarrow NaHCOO + H_2O$$

If HCl is added to the same buffer, then:

$$HCl + NaHCOO \longrightarrow NaCl + HCOOH$$

3.10: Buffer solutions

1 Explain how you would prepare a buffer of $CH_3CH_2NH_2/CH_3CH_2NH_3Cl$ and explain what would happen to the capacity of the buffer if small amounts of HCl were added (include a relevant equation).

____

____

____

____

____

____

____

 ISBN: 9780170352611

2 a Calculate the pH of a HF/NaF buffer if the concentration of HF is 0.100 $molL^{-1}$, the concentration of NaF is 0.0500 $molL^{-1}$ and $K_a(HF) = 7.2 \times 10^{-4}$.

b Calculate the pH of a CH_3COOH/KCH_3COO buffer if the concentrations of each species is the same and $K_a(CH_3COOH) = 1.76 \times 10^{-5}$.

3 Calculate the capacity factor of a $H_2C_2O_4/HC_2O_4^-$ buffer if 15 mL of it with pH = 1.23 is diluted to 50 mL. $pK_a = 1.23$

ISBN: 9780170352611

Titration curves

A titration is a way of accurately determining the concentration of a species using a species with known concentration. In this part of the unit you will need to be able to interpret, describe and calculate certain points of titration curves for acid-base reactions as well as pick out which indicator will best represent the end point of the titration.

1 Strong acid titrated with strong base

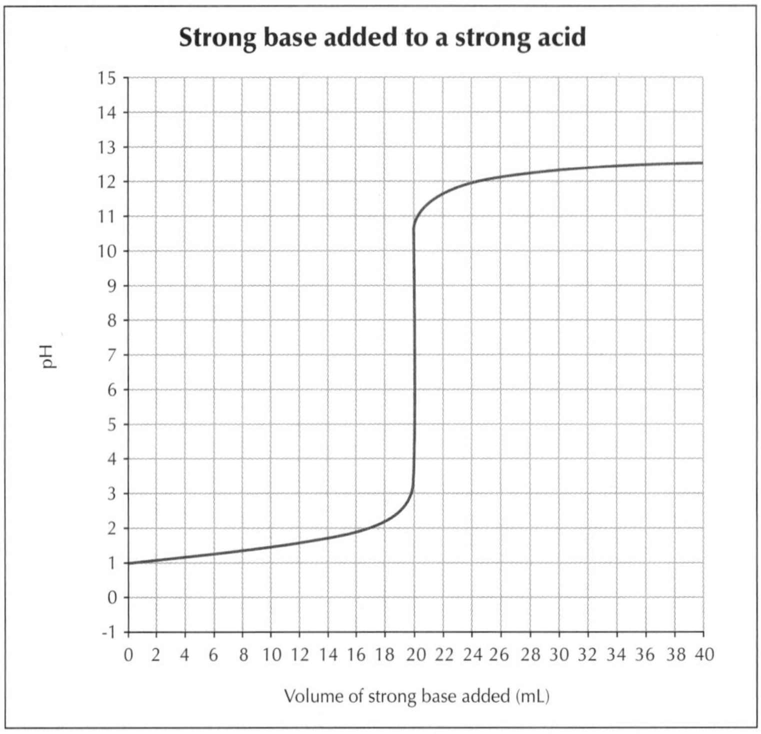

2 Strong base titrated with strong acid

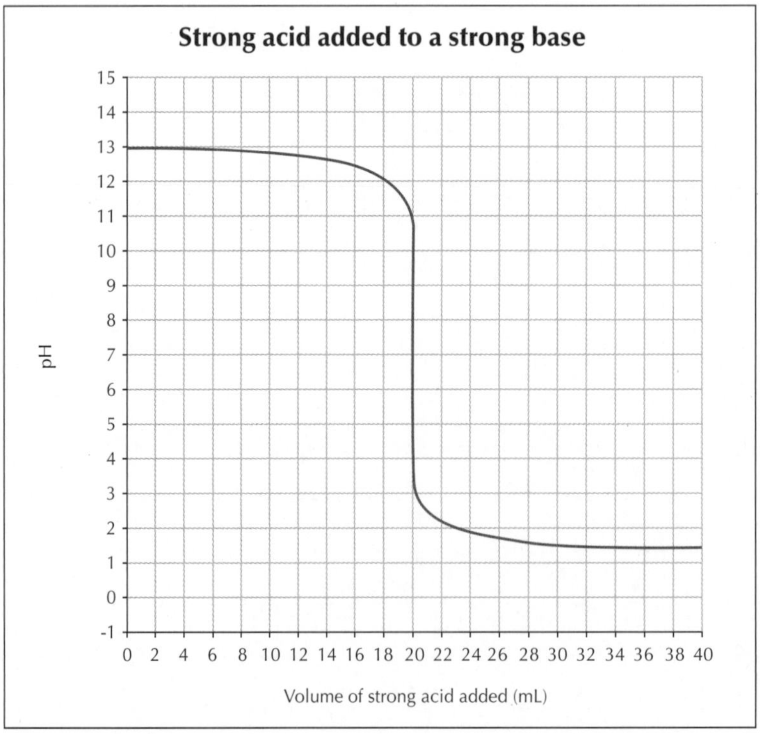

  ISBN: 9780170352611

In titrations with strong acid and strong base, the initial pH can be calculated if you know the concentration of the acid if starting from the acid or of the base if starting with the base, using the calculations we used earlier in the chapter for calculating the pH of a strong acid/base.

The equivalence point is the point where all of the acid or base has been neutralised by its counterpart. This means that the pH is always 7 at equivalence point in these curves. The straight line through this part of the graphs occurs because after this point the pH rapidly changes to the concentration of the solution that was titrated against.

The final pH will equal the pH of the strong acid or base depending on what you titrated with.

Example:

If 0.100 $molL^{-1}$ HCl is titrated with 20 mL of NaOH and 20 mL of HCl is required for the end point, then the following pHs can be calculated.

Initial pH: Here only the base is present so we need to write an equation to work out how many moles of NaOH were present.

$$NaOH + HCl \longrightarrow NaCl + H_2O$$

n(NaOH):n(HCl) is 1:1 so n(HCl) = concentration x volume = 0.1 x 20/1000 = 0.002 mol = n(NaOH)

The concentration of NaOH is then c(NaOH) = n(NaOH)/V(NaOH) = 0.002/(20/1000) = 0.100 $molL^{-1}$

$K_w = [OH^-][H_3O^+]$ $[H_3O^+] = K_w/[OH^-] = 1 \times 10^{-14}/0.1 = 1 \times 10^{-13}$ $pH = -\log[H_3O^+] = 13.0$

At equivalence point the pH is 7.
After the equivalence point, all that is left is the acid as all the base has been neutralised, so:

$[HCl] = [H_3O^+] = 0.100\ molL^{-1}$
$pH = -\log[H_3O^+] = 1.00$

3 Weak acid titrated with strong base

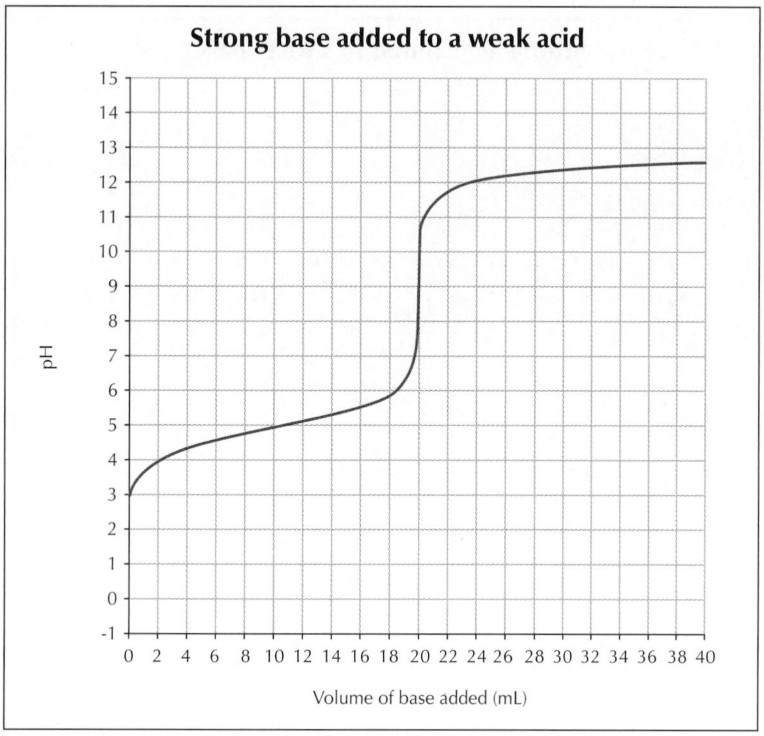

ISBN: 9780170352611

In a weak acid versus strong base titration the initial pH can be determined by calculating the pH of the weak acid as shown earlier in the chapter. Halfway to equivalence point there is a buffer zone where the pK_a of the weak acid equals the pH. This part of the curve is relatively flat as a buffer has been created and so does not alter its pH very much upon the addition of the base.

The equivalence point is always above 7 and can be calculated by working out the concentration of the conjugate base to the weak acid's concentration.

After the equivalence point the pH is solely relating to the pH of the strong base.

Example:

If 20 mL of 0.100 $molL^{-1}$ ethanoic acid, $K_a(CH_3COOH) = 1.76 \times 10^{-5}$ is titrated with potassium hydroxide, KOH, 20 mL of the base are required for the titration to reach equivalence point.

Initial pH: $[H_3O^+] = \sqrt{K_a \times [CH_3COOH]} = 0.00132...\ molL^{-1}$ $\quad$ $pH = -\log[H_3O^+] = 2.88$

Buffer zone: $pH = pK_a = -\log K_a = 4.75$

Equivalence point: $CH_3COO^- + H_2O \rightleftharpoons CH_3COOH + OH^-$

$[H_3O^+] = \sqrt{K_a \times K_w/[CH_3COO^-]}$

Since the moles of CH_3COOH = CH_3COO^- and the volume has doubled, $[CH_3COO^-] = 0.1/2 = 0.0500\ molL^{-1}$
$[H_3O^+] = \sqrt{1.76 \times 10^{-5} \times 1 \times 10^{-14}/0.05} = 1.87... \times 10^{-9}\ molL^{-1}$
$pH = -\log[H_3O^+] = 8.73$

After equivalence point: $CH_3COOH + KOH \longrightarrow KCH_3COO + H_2O$
$n(CH_3COOH):n(KOH) = 1:1$
$n(CH_3COOH) = n(KOH) = c(CH_3COOH) \times V(CH_3COOH) = 0.1 \times 20/1000 = 0.002\ mol$
$c(KOH) = n(KOH)/V(KOH) = 0.002/(20/1000) = 0.100\ molL^{-1}$
$[H_3O^+] = K_w/[OH^-] = 1 \times 10^{-14}/0.1 = 1 \times 10^{-13}\ molL^{-1}$
$pH = -\log[H_3O^+] = 13$

4 Weak base titrated with strong acid

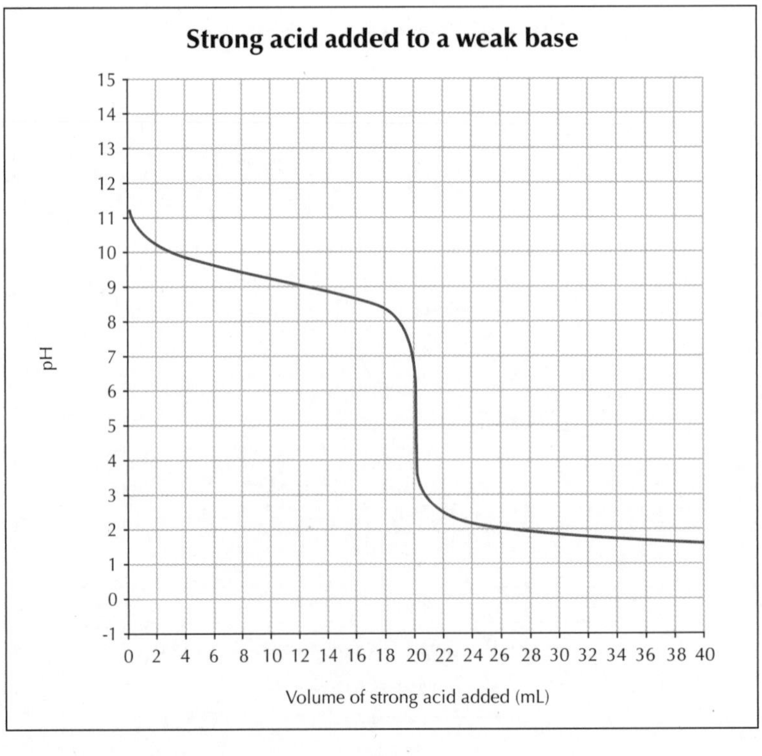

  ISBN: 9780170352611

In a weak base versus strong acid titration the initial pH relies solely upon the pH of the weak base. It too has a buffer region at halfway to equivalence point where the pK_a of its conjugate acid is equal to the pH. Its pH at equivalence point is less than 7 as all the weak base has reacted leaving only the conjugate acid in solution. Finally the pH after equivalence point relies solely on the pH of the strong acid.

3.11: Titration curves

1 a Label the points of the following titration curve of a weak acid reacting with a strong base.

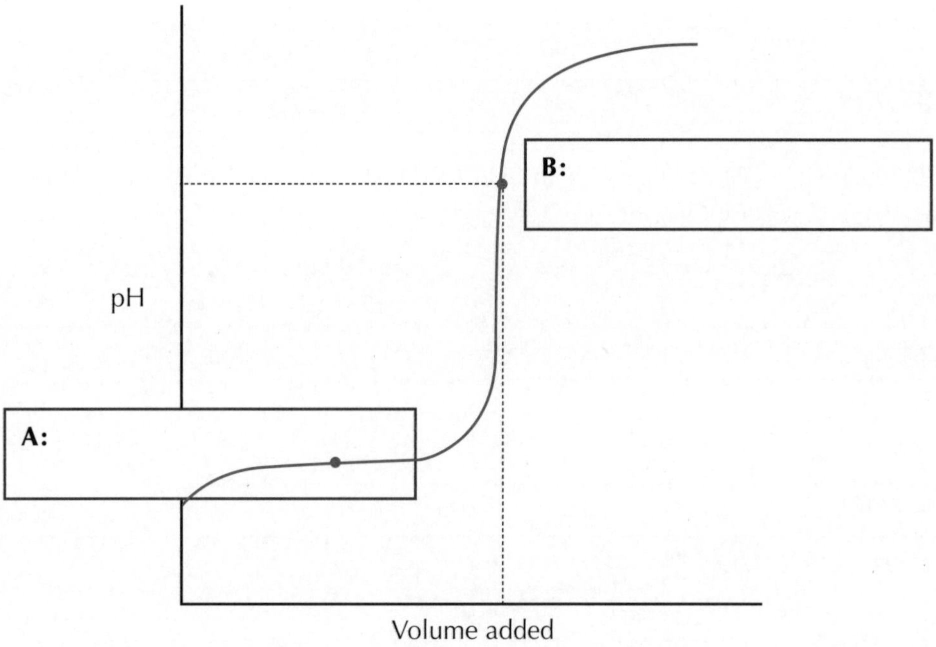

b Explain how you could determine the pH of section A and why its pH does not vary much despite the addition of more base.

2 For the following titrations calculate the initial pH, the pH at equivalence point and the final pH.

a 25 mL of 0.0150 $molL^{-1}$ NaOH was titrated with HCl; at equivalence point the volume of HCl added was 20 mL.

b 15 mL of 0.200 $molL^{-1}$ KOH was titrated with CH_3COOH; at equivalence point the volume of CH_3COOH added was 20 mL. $K_a(CH_3COOH) = 1.76 \times 10^{-5}$

c 10 mL of 0.600 $molL^{-1}$ NH_3 was titrated with HCl; at equivalence point the volume of HCl added was 12 mL. $pK_a(NH_4^+) = 9.25$

ISBN: 9780170352611

Indicators

Indicators are added in order to determine the end point of a titration.

- **Equivalence point** = the point in the titration where the correct amount of base has been added to react completely with the acid.
- **End point** = when the indicator changes colour.

Indicators are made from weak acids where the colour varies when the conjugate base is formed.

An indicator is chosen as it changes colour at close to the equivalence point so that the experimenter knows that equivalence point has been reached and all the acid or base has been neutralised. Indicators with similar pK_a values to the pH at equivalence point are chosen, with the best changing colour and with pK_a values that are slightly higher than the pH at equivalence point.

CHEMISTRY APPS

A certain variety of hydrangea changes colour depending on the pH of the soil it is in. Many other plants will also act as indicators, for example onion changes its smell depending on the pH of the solution it is in.

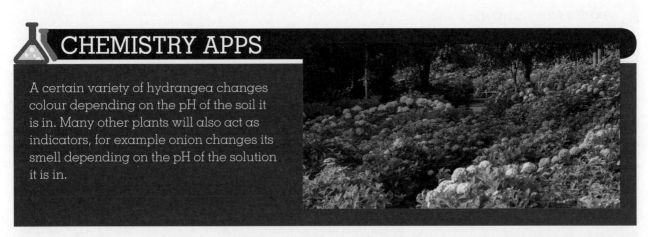

3.12: Indicators

Choose which indicator from the table below will work best with the following titrations and explain why you chose it.

Indicator	pK_a
Bromothymol blue	7.1
Phenolphthalein	9.3
Methyl red	5.1

a A titration with HCl and KOH:

b A titration with HCl and NH_3:

c A titration with HCOOH and KOH:

ISBN: 9780170352611

EXPERIMENT 3

Which indicator is best?

AIM: To investigate the colour certain indicators change in different pH solutions.

EQUIPMENT:

methyl orange
2 $molL^{-1}$ sodium hydroxide
red cabbage
litmus
phenolphthalein
universal indicator
mortar and pestle
2 $molL^{-1}$ ammonia
2 $molL^{-1}$ hydrochloric acid
bromothymol blue
methyl red
2 $molL^{-1}$ ethanoic acid

SAFETY PRECAUTIONS:

Find out about the safety issues for this lab, by looking up the material safety data sheet (MSDS) for each species.

METHOD:

Make up your own method for testing the two acids (HCl and CH_3COOH) and two bases (NH_3 and NaOH). Make sure your method can be followed by someone else.

CONCLUSION:

Which indicator(s) worked best? Why?

 ISBN: 9780170352611

What do the following key words mean?

Dissociate	
Acid	
Base	
pH	
K_a, acid dissociation constant	
Buffer	
Titration	
End point	
Equivalence point	
Indicator	

CHECKPOINT 3
WHAT HAVE YOU LEARNED SO FAR?

Buffers, pH and conductivity of acids and bases

1 Calculate the pH of the following acids or bases.

a 0.150 $molL^{-1}$ HCl

b 0.200 $molL^{-1}$ CH_3COOH, $K_a = 1.76 \times 10^{-5}$

c 0.608 $molL^{-1}$ KOH

d 0.112 $molL^{-1}$ CH_3NH_2, $pK_a(CH_3NH_3^+) = 10.7$

e Explain which solution out of 0.100 $molL^{-1}$ KOH and 0.100 $molL^{-1}$ CH_3NH_2 above will conduct the best.

2 Explain how a buffer solution could be made from 0.100 $molL^{-1}$ CH_3COOH and 0.150 $molL^{-1}$ KCH_3COO; include a calculation of its pH in your answer. $K_a(CH_3COOH) = 1.76 \times 10^{-5}$

ISBN: 9780170352611

CHECKPOINT 3
WHAT HAVE YOU LEARNED SO FAR?

3 Explain how you would calculate the initial pH, the pH of the buffer zone, the pH at equivalence point and the final pH for a titration with ammonia, NH_3 and nitric acid, HNO_3. No calculations are required here.

ISBN: 9780170352611

EXAM-TYPE QUESTIONS

Before you get started here are some points to remember from the Examiner's Report:

- Don't use generalisations; use terms carefully.
- Plan your answer before you start.

Question one

Our teeth have a hard enamel coating, $Ca_5(PO_4)_3OH$, which protects them. This enamel is made in the following equilibrium:

$$5Ca^{2+}_{(aq)} + 3PO_4^{3-}{}_{(aq)} + OH^-_{(aq)} \rightleftharpoons Ca_5(PO_4)_3OH_{(s)}$$

Sugary foods produce weak acids such as ethanoic acid (CH_3COOH) and lactic acid ($CH_3CH(OH)COOH$) in our mouths. Discuss what would happen to the solubility of enamel coating if you ate too much sugary food and what happens to its solubility when you brush your teeth.

ISBN: 9780170352611

EXAM-TYPE QUESTIONS

Question two

The sparingly soluble substance $Be(OH)_2$ has a $K_s = 6.92 \times 10^{-22}$ at 25°C.

a Calculate its solubility in distilled water.

b Discuss the effect on its solubility if the pH was raised to 11.5.

ISBN: 9780170352611

Question three

Zinc hydroxide, $Zn(OH)_2$, is sparingly soluble. $K_s(Zn(OH)_2) = 3 \times 10^{-17}$

Discuss the effect on its solubility if the pH is changed to the following values (include equations and calculate values).

a pH = 9.2

b pH = 4.5

c Discuss the effect of ammonia, NH_3, being added to the equilibrium mixture.

 ISBN: 9780170352611

EXAM-TYPE QUESTIONS

Question four

a What mass of potassium chloride must be added to 1 L of 0.100 $molL^{-1}$ silver nitrate solution to form a precipitate? $K_s(AgCl) = 2 \times 10^{-10}$; $M_r(KCl) = 74.6\ gmol^{-1}$

b Discuss the effect of adding ammonia, NH_3, to an equilibrium mixture of silver chloride.

Question five

a What is the pH of a solution containing 0.501 molL^{-1} of HSO_4^-? $pK_a(HSO_4^-) = 6.91$

b Calculate the pH of HSO_4^- conjugate base SO_4^{2-} if there is a solution of 0.166 molL^{-1} SO_4^{2-}.

ISBN: 9780170352611

EXAM-TYPE QUESTIONS

Question six

Discuss the relative concentrations of all species, except water, in 0.100 $mol L^{-1}$ solutions of $CH_3CH_2NH_2$ and KHCOO. Include relevant equations and rank the species in decreasing concentration in each solution. Justify your rankings.

ISBN: 9780170352611

Question seven

The pH and conductivity of some solutions were measured and recorded in the table.

Solution	pH	Conductivity
KCl	7	High
CH_3CH_2COOH	4.06	Low
CH_3CH_2COONa	8.91	High

Discuss the data in terms of species present in the solutions.

ISBN: 9780170352611

EXAM-TYPE QUESTIONS

Question eight

a Calculate the pH of a buffer solution made with equal amounts of ammonia, NH_3 and ammonium chloride, NH_4Cl. $K_a(NH_4Cl) = 5.6 \times 10^{-10}$

b Explain what would happen to the buffer solution upon the addition of small amounts of hydrochloric acid, HCl.

Question nine

Calculate the data points to plot a pH curve for the titration of 12 mL of 0.500 $molL^{-1}$ ethanoic acid with 0.500 $molL^{-1}$ NaOH. $K_a(CH_3COOH) = 1.76 \times 10^{-5}$

ISBN: 9780170352611

EXAM-TYPE QUESTIONS

Question ten

A titration was carried out using 15 mL of 0.0150 $molL^{-1}$ sodium benzoate, NaC_6H_5COO with 15 mL of HCl being used at equivalence point. Calculate the initial pH, the buffer region pH, the pH at equivalence point and the final pH for this curve. $K_a(C_6H_5COOH) = 6.46 \times 10^{-5}$

INTERNAL STANDARDS REVIEW

ISBN: 9780170352611

3.1 — Carry out an investigation in chemistry involving quantitative analysis (91387)

4 credits

Here are the main things you need to know in order to get the grade you desire:

Achieved	In this investigation you must come up with a **purpose**, **method** (which includes preparing of samples that you will test from and any analytical test used), collect, record and process all **data** in a **logbook** and a **conclusion** based on what you found out. You may collect data in groups but must hand in *your own* report and logbook. Your **purpose** is the aim of the investigation and it must be related to a trend or pattern in the quantity of substance in a sample. For example: The purpose of this investigation is to measure the quantity of vitamin C in a variety of brands of orange juice and see how it relates to the nutritional information found on the packaging.  Your **method** must include step-by-step instructions of how you will measure this quantity, what equipment you will use, and must include **four** values for the independent variable (for example, four different brands of orange juice). *This must be as detailed as possible and written for someone who has no idea what you are doing.* Your **results** will include all data you collect from your analytical technique (for example, titre volumes), and any calculations you complete in order to make sense of this data (for example, the concentration of vitamin C). Your **conclusion** should then link your independent variable to your dependent variable. For example: The nutritional information on each of the four packets matched closely with the amount of vitamin C found in each sample.
Merit	At Merit level you need more detail at each stage of the investigation of what was done, how it was carried out and what you found out. Your **method** must include a description of how you made a standard solution and how you controlled all other variables. For example: I collected all my data using the same equipment and used the same standard solution to measure the amount of vitamin C in the different orange juices. Your method also must be detailed enough that someone else could replicate your results and this time have **five** values for your independent variable. It should also outline any safety issues and any changes that were made during the investigation. Your **results** section must include all working for the calculations you complete and process your data correctly in order to reach a valid conclusion. You must also show concordant results if completing a titration.

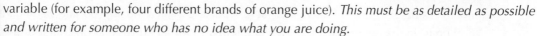

Merit	Your **conclusion** should link back to your purpose. For example: In this investigation it was found that the nutritional information given on the packaging of five different brands of orange juice was similar to the quantities found when this investigation was completed. It should also include some quantitative value. For example: Brand X contained 30% more vitamin C than brand Y. There is the addition of an **evaluation** at this level. This must include an explanation of how the method you used provided you with quality data. For example: The method I used contained very little experimental error and is the same technique used by the orange juice companies to test their vitamin C quantities. Finally, include a bibliography where all resources are presented in such a way that they can be easily found by someone else.
Excellence	At Excellence level all of the requirements at Achieved and Merit level are needed with a few more detailed accounts of what you did and found out from your investigation. In the **method** section be sure to justify any changes made. You would make changes if, for example, you were getting unusual titre values that were too low or too high after some initial trials. You should also repeat the entire process from start to finish using a different standard solution in order to be sure your results are reliable, especially given the nature of a chemical such as vitamin C and how long it might last. In the **results** section you must make sure all data contains the appropriate significant figures and units (for example, use three significant figures for all concentrations but four for all titre volumes). Your results should be within the level of accuracy that is possible for you using the equipment and chemicals provided. In the **conclusion**, link all findings to chemical principles and/or real-life applications. For example: Brands need to state accurate information on their packaging as part of the requirements of selling a product in New Zealand and so they can guarantee their goods to the consumer. In the **evaluation** you must justify the steps you used in your method and how they were chosen. You must also evaluate the reliability of your data by looking at sources of error and how the data supports your conclusion. This also includes a discussion of how relevant you think your findings are: would they change consumers' minds in the vitamin C case, or are the results as expected? For example: The main source of error in this experiment came from the glassware, as both the pipette and burette contained ±0.1 mL error.

ISBN: 9780170352611

3.2 — Demonstrate understanding of spectroscopic data in chemistry (91388)

3 credits

Here are the main things you need to know in order to get the grade you desire:

There are three different types of spectroscopic data that will be given to you in order for you to pinpoint the organic molecule shown in the data: ***mass***, ***IR*** and ***^{13}CNMR spectroscopy***. (You do not need to explain how these methods work; you just need to be able to interpret them. Below are a few key features of each.)

Mass spectrometry

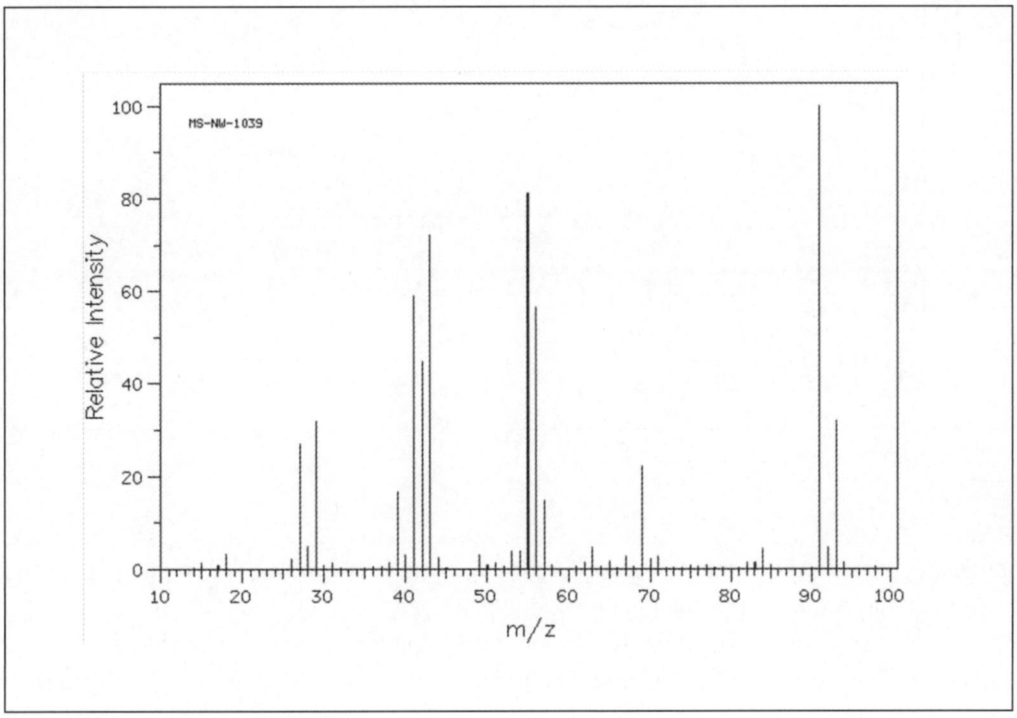

The Mass Spectrum of 1-chlorohexane

This technique provides you with the mass of the molecular ion (mass spectrometry turns the molecule in question into an ion); this ion is unstable and so splits off into smaller ions known as fragments. These fragments can help identify the molecule that is present. Note that sometimes the molecular ion cannot be identified from a spectrum as the ion is too unstable to be measured. In the example above, the molecular ion peak is not significant and so is not shown, however the fragment at 93 shows a loss of 29 giving us the fragment $C_4H_8Cl^+$.

IR spectroscopy

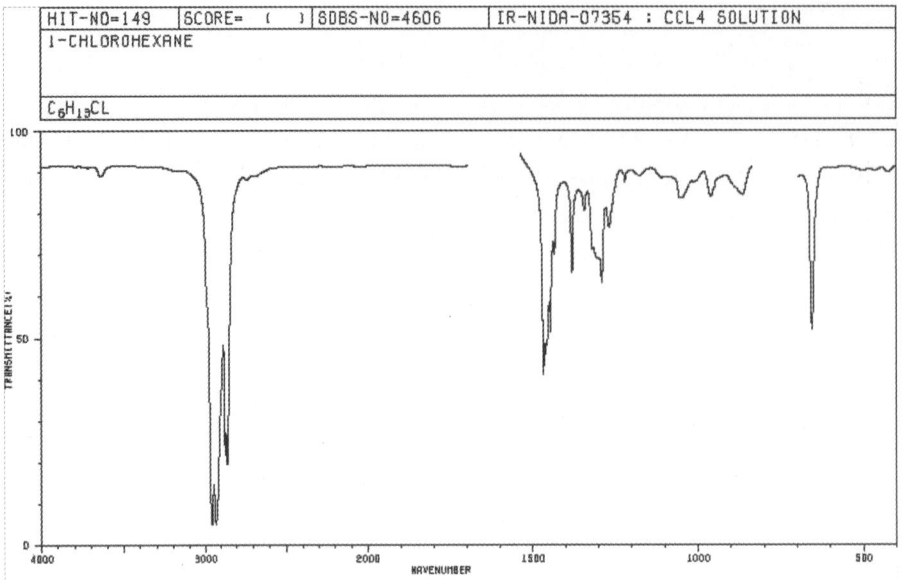

3638	86	1467	47	1290	60	970	84
2959	4	1446	50	1268	74	961	81
2933	4	1433	68	1221	84	664	81
2874	21	1380	64	1181	86	666	60
2861	19	1342	77	1175	86		
1468	39	1302	66	1054	81		
1462	44	1297	68	1046	81		

IR of 1-chlorohexane

IR shows bands that relate to the functional group or groups present in a molecule. The large peak at 2800 cm^{-1} will be present in all molecules as it is the C-H peak; the peak between 600 and 800 cm^{-1} could be a C-Cl peak, however it is hard to determine exactly what is present in this region due to the presence of the solvent which was used. Other functional groups will also contain bands which you can identify using the data table that is provided for you.

^{13}CNMR spectroscopy

This technique shows peaks for each different carbon environment present.

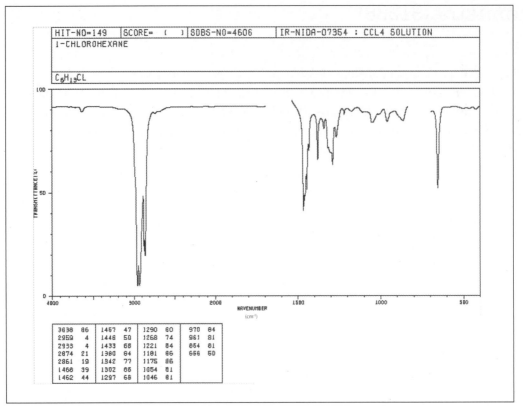

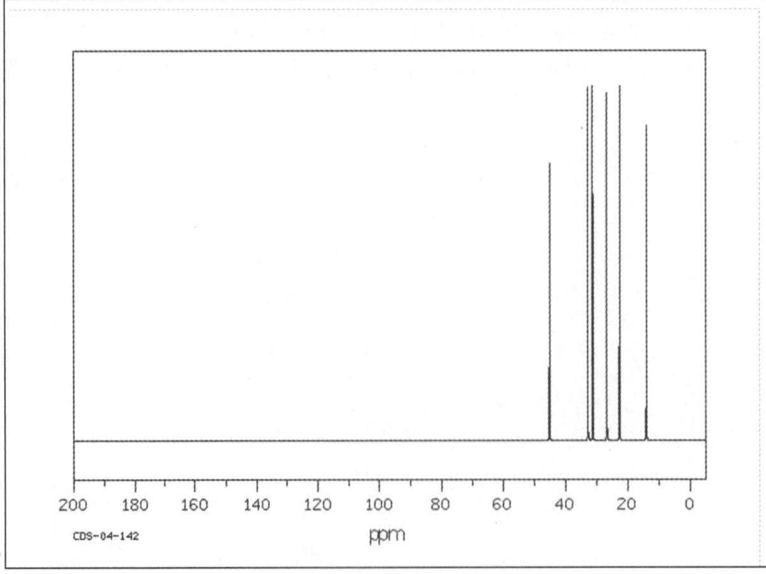

^{13}CNMR for 1-chlorohexane

 ISBN: 9780170352611

In this example there are six peaks, as all six carbons are in slightly different environments. There is also a distinct shift at 45 ppm which is the carbon connected to the Cl.

Achieved	At Achieved level you must interpret a few features from some spectroscopic data that would help to identify the organic molecule. In the example above, identifying one peak in the mass spectrum and linking with the fragment would be one feature that could be used to get you an Achieved. Or you could identify one peak in the IR spectrum. Finally, using the ^{13}CNMR spectrum you could identify how many different environments for carbon there are; in this case there would be six, as this molecule is not symmetrical. The organic molecules you could be asked to identify come from the following groups: *alkanes, alkenes, alcohols, haloalkanes, amines, aldehydes, ketones, carboxylic acids, amides, acid chlorides* and *esters.*
Merit	At Merit you must identify the organic molecule shown from the spectroscopic data. So in this example you would identify that the molecule is 1-chlorohexane and link it to the peaks you located in the spectra.
Excellence	At Excellence level you must justify your choice of molecule using the spectroscopic data. At Excellence level you need to link up the organic molecule with information from all three spectra as outlined above, so that no other molecule could be interpreted from these peaks.

3.3 — Demonstrate understanding of chemical processes in the world around us (91389)

3 credits

Here are the main things you need to know in order to get the grade you desire:

Achieved	At Achieved level you need to give an account of a chemical process that occurs in the natural world around us or that solves a problem or a need. This will include any equations that may occur as well as a brief description of the chemistry involved in this process. For example: *If you are writing on the topic of greenhouse gases you may include the combustion of fuel equation forming carbon dioxide and explain other sources of where the gases come from, along with an explanation of what the effect of greenhouse gases are.* 
Merit	At Merit level you need to explain this chemical process and how it affects either the natural world or how it solves a problem or need. For example: *The explanation would include where the greenhouse gases have come from and possible changes that could be made to lower how much of them are produced. Also you would talk about the effects of these greenhouse gases.*
Excellence	At Excellence level you need to evaluate this process in terms of the impact of this issue or problems that have come from this issue; as well as more depth at every level of your answer with constant links back to the chemistry involved. For example: *The word evaluate is crucial here so use it to explain and justify where these greenhouse gases have come from, why they are an issue, possible ways to solve the issue as well as a detailed discussion of how greenhouse gases are altering the planet's atmosphere.*

Possible chemical processes or issues you could write about include: the greenhouse effect, ozone depletion, acidification of the oceans, acid rain, volcanic eruptions, pollution, polymers, energy production, pharmaceuticals or food production.

ISBN: 9780170352611

3.7 — Demonstrate understanding of oxidation-reduction processes (91393)

3 credits

Here are the main things you need to know in order to get the grade you desire:

<table>
<tr><td>Achieved</td><td>At Achieved level you will need to be able to label electrochemical cells and electrolytic cells, correctly deduce the direction of electron flow, identify which species is reduced and which is oxidised and calculate E°cell potentials. For example:

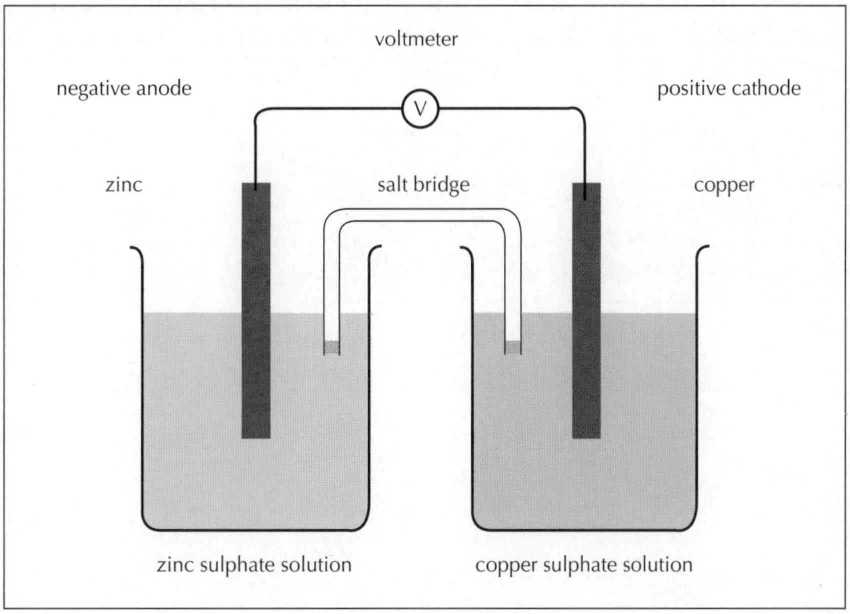

In the above example, all the labels have been given to you except the flow of electrons. The value of the copper half-cell is +0.34 V and the zinc half-cell is -0.76 V (this means that copper is a stronger oxidant and so it will be reduced). Therefore electrons will flow from the left to the right half cell.
E°cell = 0.34 – -0.76 = 1.1 V</td></tr>
<tr><td>Merit</td><td>At Merit level you will need to link the reaction occurring in each half-cell to a half-equation, a cell diagram, explain the flow of electrons, explain a cell in terms of its E°cell potential and include a balanced equation. For example:

In the left half-cell at the anode, the following reaction will occur:
$Zn \longrightarrow Zn^{2+} + 2e^-$
In the right half-cell at the cathode, the following reaction will occur:
$Cu^{2+} + 2e^- \longrightarrow Cu$
The oxidation-reduction equation for this reaction therefore will be:
$Zn + Cu^{2+} \longrightarrow Zn^{2+} + Cu$
The electrons will flow from the oxidation reaction at the anode to the reduction reaction at the cathode; the circuit is completed by the flow of charge through the salt bridge.
The cell diagram for this cell would be:
$Zn_{(s)} \mid Zn^{2+}_{(aq)} \mid\mid Cu^{2+}_{(aq)} \mid Cu_{(s)}$</td></tr>
</table>

ISBN: 9780170352611

Excellence	At Excellence level you will need to be able to explain all aspects of the cell including how the flow of charge completes the circuit. You should also be able to explain the differences between an electrochemical and an electrolytic cell. For example: Within the salt bridge a solution of KNO_3 may be added where the K^+ ions move towards the negative cathode in order to replace the charge that is lost from the reduction and the NO_3^- ions move towards the positive anode in order to replace the charge that is lost from the oxidation reaction. This flow of charge allows the current to flow and completes the circuit. Electrochemical cells do not require energy for the spontaneous reaction to occur and the anode is negative and the cathode is positive due to the oxidation-reduction reactions occurring. Electrolysis requires energy for the reactions to occur as the oxidation-reduction processes occurring are not spontaneous. The anode is positive and the cathode is negative due to the current that is provided from the power source which forces the reactions to occur.

ISBN: 9780170352611

PRACTICE ASSESSMENT SECTION

Level 3 Chemistry

91390 Demonstrate understanding of thermochemical principles and the properties of particles and substances

3

Credits: Five

Achievement	Achievement with Merit	Achievement with Excellence
Demonstrate understanding of thermochemical principles and the properties of particles and substances.	Demonstrate in-depth understanding of thermochemical principles and the properties of particles and substances.	Demonstrate comprehensive understanding of thermochemical principles and the properties of particles and substances.

You should attempt ALL the questions in this booklet.

You are advised to spend 60 minutes completing this.

TOTAL

QUESTION ONE

TEACHER USE ONLY

a Write the electron configuration for the following atoms or ions.

Sc

Zn^{2+}

b i Discuss the difference in the atomic and ionic radii of Zn shown in the table below.

Zn atomic radii (pm)	135
Zn^{2+} ionic radii (pm)	74

ISBN: 9780170352611

TEACHER USE ONLY

ii Discuss the change in ionisation energies of zinc shown in the table below.

First ionisation energy ($kJmol^{-1}$)	906.4
Second ionisation energy ($kJmol^{-1}$)	1733
Third ionisation energy ($kJmol^{-1}$)	3833

c i Fill in the table below of the Lewis diagram and shape of the following molecules.

	ClF_3	ICl_4^-
Lewis diagram		
Shape		

ii Compare and contrast the polarity and shapes of these two molecules.

TEACHER USE ONLY

QUESTION TWO

a i Calculate the $\Delta_r H$ for the following reaction:

$$CS_{2\,(l)} + 2O_{2\,(g)} \longrightarrow CO_{2\,(g)} + 2SO_{2\,(g)}$$

given the following:

$\Delta_f H^o(CO_{2\,(g)}) = -393.5\ kJmol^{-1}$

$\Delta_f H^o(SO_{2\,(g)}) = -296.8\ kJmol^{-1}$

$\Delta_f H^o(CS_{2\,(l)}) = 87.9\ kJmol^{-1}$

ISBN: 9780170352611

ii Explain what $\Delta_f H^o(CO_{2\,(g)})$ represents.

TEACHER USE ONLY

b i Calculate the mass when nitric acid is mixed with magnesium hydroxide solution, which had a change in temperature of 6.25°C and it released 3500 J of energy.

$c\,(H_2O) = 4.18\ J^oC^{-1}g^{-1}$

ii Explain why a value that is calculated in a school laboratory for the reaction above may be lower than a theoretical value.

ISBN: 9780170352611

TEACHER USE ONLY

QUESTION THREE

a **i** Explain what $\Delta_{sub}H^o$ means given that $\Delta_{sub}H^o(I_2) = 62.0$ kJmol^{-1}.

ii Explain why the $\Delta_{sub}H^o$ of iodine, I_2, is so high. Include both enthalpy and entropy changes in your answer.

ISBN: 9780170352611

TEACHER USE ONLY

b Compare and contrast the electronegativity values as you go down group 17.

Halogen	Electronegativity value
F	4.0
Cl	3.0
Br	2.8
I	2.5

c Compare and contrast the boiling points of the following molecules.

Substance	Boiling point (°C)
Iodine trichloride, ICl_3	33.0
Ethanoic acid, CH_3COOH	118
Octanoic acid, $CH_3(CH_2)_6COOH$	237

TEACHER USE ONLY

ISBN: 9780170352611

Level 3 Chemistry

91391 Demonstrate understanding of the properties of organic compounds

3

Credits: Five

Achievement	Achievement with Merit	Achievement with Excellence
Demonstrate understanding of the properties of organic compounds.	Demonstrate in-depth understanding of the properties of organic compounds.	Demonstrate comprehensive understanding of the properties of organic compounds.

You should attempt ALL the questions in this booklet.

You are advised to spend 60 minutes completing this.

TOTAL

QUESTION ONE

TEACHER USE ONLY

a Give the IUPAC name or draw the structure for the following compounds.

IUPAC name	Structure
ethyl methanoate	
	CH_3CH_2COOH
ethanal	
	CH_3CH_2COCl
butan-2-ol	

ISBN: 9780170352611

b Explain how you could distinguish between CH_3CH_2COOH and CH_3CH_2COCl by their boiling point.

In your answer you should include:

- the type of intermolecular bonds present between the molecules of these two compounds
- the relative energy required to break these intermolecular bonds
- an explanation of how you could distinguish between these two compounds using boiling point alone.

TEACHER USE ONLY

ISBN: 9780170352611

c Butan-2-ol can form enantiomers. Explain what an enantiomer is and what is the requirement for enantiomers to form as well as a physical test that could be used to distinguish between these two enantiomers. In your answer include three-dimensional drawings of both of these enantiomers.

TEACHER USE ONLY

TEACHER USE ONLY

QUESTION TWO

a Complete the reaction scheme below.

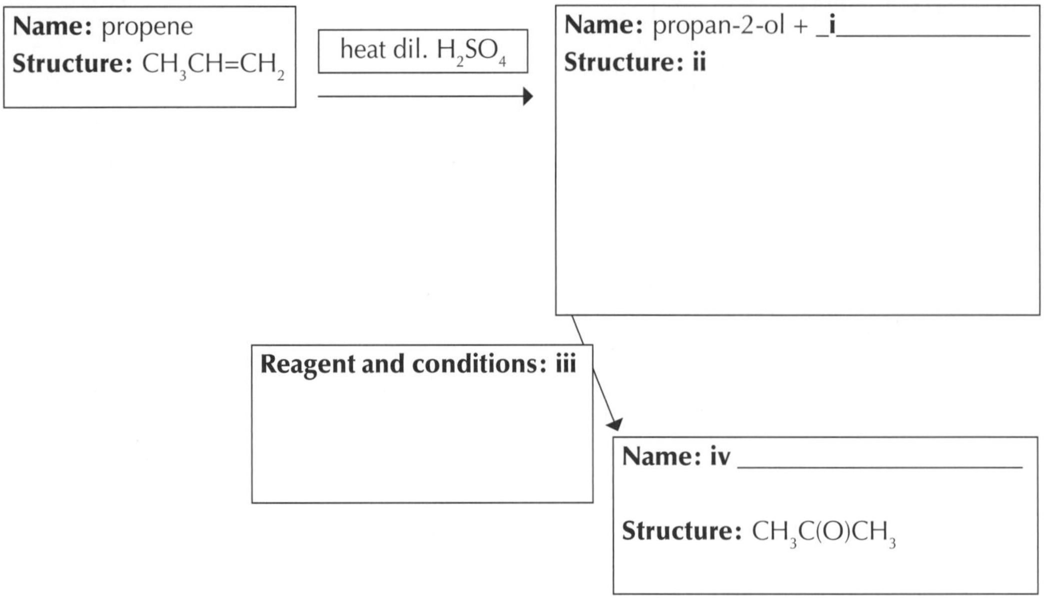

b Complete the reaction scheme below and explain the types of reactions occurring in each, how they are similar and how they are different.

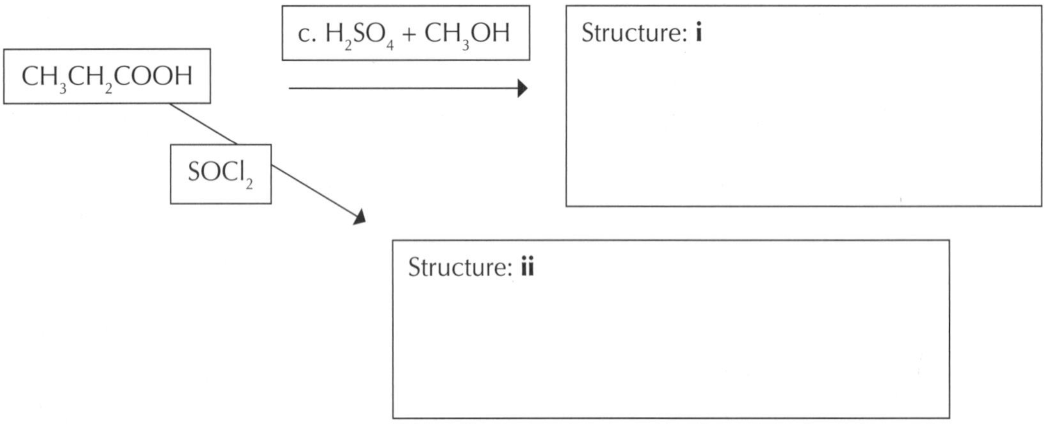

Explanation of the type of reactions occurring and why they are different.

ISBN: 9780170352611

QUESTION THREE

TEACHER USE ONLY

a Complete the two hydrolysis reactions shown below.

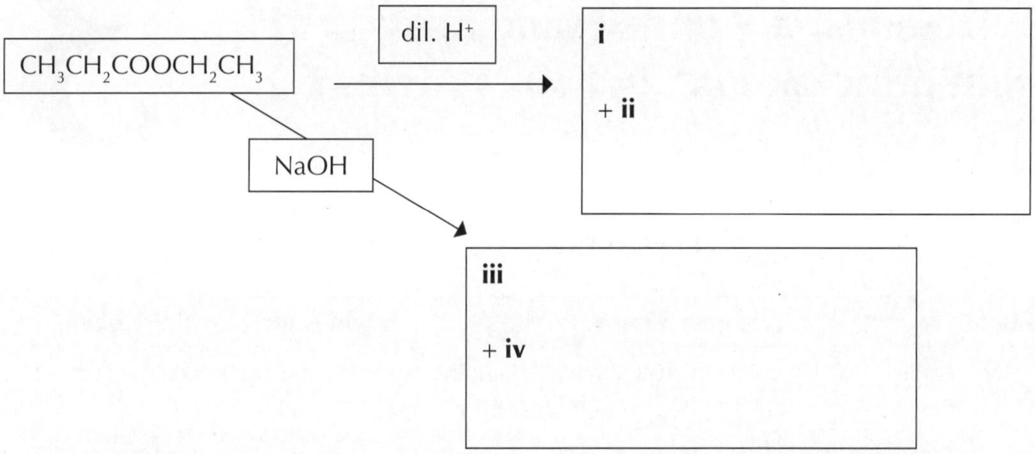

b Complete the condensation polymerisation reaction shown below by drawing in two repeating units of the polymer product.

$HOOCCH_2COOH + HOCH_2OH \longrightarrow$

c Compare and contrast the two reactions shown above, explaining why they are different reactions forming different products.

Level 3 Chemistry

91392 Demonstrate understanding of equilibrium principles in aqueous systems

3

Credits: Five

Achievement	Achievement with Merit	Achievement with Excellence
Demonstrate understanding of equilibrium principles in aqueous systems.	Demonstrate in-depth understanding of equilibrium principles in aqueous systems.	Demonstrate comprehensive understanding of equilibrium principles in aqueous systems.

You should attempt ALL the questions in this booklet.

You are advised to spend 60 minutes completing this.

TOTAL

QUESTION ONE

TEACHER USE ONLY

a Rank the species present in the following solutions from highest to lowest.

i HNO_3

ii CH_3COOH

iii $NaCH_3COO$

b Compare and contrast the conductivity of 1 $molL^{-1}$ of each of the above solutions.

 ISBN: 9780170352611

c Explain how a buffer solution could be made with two of the above solutions and what would happen to this buffer solution if small amounts of base were added to the buffer.

No calculations are required in this answer.

TEACHER USE ONLY

QUESTION TWO

a The K_s for zinc fluoride, ZnF_2, is 3.04×10^{-2}.

i Write the K_s expression for zinc fluoride.

$K_s =$

ISBN: 9780170352611

TEACHER USE ONLY

ii Calculate the solubility of zinc fluoride in $molL^{-1}$.

iii Calculate the solubility of zinc fluoride in a solution of 0.100 $molL^{-1}$ sodium fluoride solution.

b Explain what would happen to the solubility of zinc fluoride if excess sodium hydroxide, NaOH, is added to it.

ISBN: 9780170352611

TEACHER USE ONLY

QUESTION THREE

a A titration was carried out with 0.100 $molL^{-1}$ ethan-1-amine, $CH_3CH_2NH_2$ and 0.100$molL^{-1}$ HCl.

Calculate the initial pH of this titration or the pH of ethan-1-amine, given the pK_a of its conjugate acid $CH_3CH_2NH_3^+$ is 10.7.

b Calculate the pH at equivalence point for this titration.

ISBN: 9780170352611

c Outline which indicator would be the best choice for this titration given the table of indicators below.

Indicator	pK_a of indicator
Methyl red	5.1
Bromothymol blue	7.0
Phenolphthalein	9.6

TEACHER USE ONLY

ISBN: 9780170352611

GLOSSARY OF TERMS

Acid: a proton-donating substance.

Acid chloride (acyl chloride): a group of organic molecules that contain a COCl functional group.

Acid dissociation constant, K_a: this a constant value under standard conditions of how much of an acid ionises in water.

Addition polymer: a type of large organic molecule made from many alkene monomers.

Alcohol: a group of organic molecules that have an OH functional group.

Aldehyde: a group of organic molecules that contain a carbonyl functional group on the end.

Alkane: a group of organic molecules that have no functional group and are made up of carbons and hydrogens singly bonded.

Alkene: a group of organic molecules that contain a C double bonded to another C as their functional group.

Amide: a group of organic molecules that contain a $CONH_2$ functional group.

Amine: a group of organic molecules containing an NH_2 functional group.

Amino acid: a type of organic molecule found in all living things that make up proteins.

Atomic radius: the distance between two atomic nuclei divided by 2.

Base: a proton-accepting substance.

Buffer: a substance made from a weak acid and its conjugate base that can maintain a stable pH.

Carbonyl: a C double bonded to an O.

Carboxylic acid: a group of organic molecules that contain COOH as their functional group.

Complex ion: an ion with ligands present in it.

Condensation polymer: a type of polymer formed when a small molecule is lost from the monomers.

Conjugate acid or base: the product from a reaction of water with either a weak base or a weak acid.

Covalent bond: a sharing of a pair or pairs of electrons between two non-metal atoms.

Dipole: a charge separation caused by electrons in a bond surrounding one atom in the bond more than the other.

Dissociate: when a substance ionises.

Distillation: a separation technique to separate out liquids, where a liquid is boiled off at its exact boiling point.

Effective nuclear charge: the attraction between the protons in the nucleus of an atom and its valence electrons.

Electron: the negative subatomic particle of an atom.

Electronegativity: the ability for an atom to attract a pair of bonding electrons towards itself.

Elimination: a type of organic reaction where a small molecule is removed from a molecule.

Enantiomer: *see* **Optical isomer**

End point: the point in a titration where an indicator changes colour.

Enthalpy: the heat content of a substance.

Enthalpy of combustion: the amount of energy required to burn one mole of a substance in oxygen.

Enthalpy of formation: the amount of energy required to make one mole of a substance from its elements in their standard states.

Enthalpy of fusion: the amount of energy required to turn one mole of substance from a solid into a liquid.

Enthalpy of sublimation: the amount of energy required to turn one mole of a substance from a solid to a gas.

Enthalpy of vaporisation: the amount of energy required to turn one mole of a liquid substance into a gas.

Entropy: the amount of disorder in a system.

Equivalence point: the point in a titration where all of the starting substance has been neutralised by the added substance or when the reaction is complete.

Ester: a group of organic molecules that contain the COO functional group.

Esterification: a condensation reaction where water or hydrochloric acid is lost in order to make an ester.

Functional group: the part of an organic molecule that is reactive.

Geometric isomer: a type of structural isomer that occurs in an alkene because there is no free rotation about the double bond and there are two different atoms or groups of atoms attached to each carbon in the double bond.

Haloalkane: an organic molecule containing the functional group of one or more halogens.

Hydrocarbon: an organic molecule made from carbons and hydrogens.

Hydrogen bond: a bond between N, F or O and H that occurs between molecules and has a special stability.

Hydrolysis: a type of reaction where water breaks apart the molecule.

Indicator: a substance that changes colour depending on how acidic or basic a solution that it is put into is.

Insoluble: when an ionic solid does not dissolve in a solvent.

Instantaneous dipole-induced dipole bond: a brief charge separation set up between two molecules forming an electrostatic interaction.

Ion: an atom or group of atoms that have lost or gained electrons.

Ionic bond: an electrostatic attraction between two oppositely charged ions.

Ionic product, Q_s: the product of the ions present in a solution that are not at equilibrium.

Ionic radius: the distance between two ionic nuclei divided by 2.

Ionisation energy: the amount of energy to remove one mole of electrons from one mole of an element in gaseous state.

Ketone: a group of organic molecules that contain a carbonyl group on one of the middle carbons.

Ligand: an attachment to a complex ion that may be an ion, atom or molecule.

Metallic bond: an electrostatic attraction between metal nuclei and their delocalised valence electrons.

Monomer: a repeating unit of a polymer.

Non-polar bond: a covalent bond where the electrons are shared equally.

Optical isomer (enantiomer): occurs when a molecule has an asymmetric carbon (a carbon with four different groups attached to it).

Orbital: the region of space where an electron may be located.

Oxidation: a type of reaction where electrons are lost.

Permanent dipole-dipole bond: a charge separation set up between two molecules forming an electrostatic interaction.

pH: a measure of how many hydronium ions are present in a solution.

Plane polarised light: light travelling in one plane only.

Polar bond: a covalent bond where the electrons are shared unequally.

Polymer: a large organic molecule made from many repeating units called monomers.

Precipitate: a solid formed in a solution.

Protein: a large organic molecule made from amino acids linked together.

Quantum level: a discrete amount of·energy that would be required to move an electron from or to an area that is a certain distance from the nucleus.

Reflux: a gentle heating method used to contain volatile reactants or products from a reaction.

Saponification: a type of hydrolysis reaction that makes a soap.

Soluble: how easily a solid dissolves in a solvent.

Solubility: how much of a solid dissolves in a solvent at a given temperature.

Solubility product, K_s: the product of the ions that dissolve in a solvent.

Solvent: a liquid that dissolves solids.

Sparingly soluble: a solid that partially dissolves in a solvent.

Species: an ion, atom or molecule that is present in a solution.

Specific heat capacity: the amount of energy (in joules) to heat 1 gram of a pure substance by 1°C.

Strong acid: an acid that fully dissociates in water to form hydronium ions and a neutral ion.

Strong base: a base that fully dissociates in water to form hydroxide ions and a neutral ion.

Structural formula: the arrangement of the atoms in an organic molecule in space.

Structural isomer: molecules with the same molecular formula but a different arrangement in space of the atoms.

Substitution: a type of organic reaction where a group of atoms or an atom are exchanged for another.

Titration: a technique used to determine the exact concentration of a substance using the concentration of a known substance.

Water dissociation constant, K_w: the amount pure water dissociates, which is equal to 1×10^{-14}.

Weak acid: an acid that only partially dissociates in water to produce hydronium ions and its conjugate base.

Weak base: a base that only partially dissociates in water to produce hydroxide ions and its conjugate acid.

Zwitterion: an ion with both acidic and basic properties.

ISBN: 9780170352611

PERIODIC TABLE

Key:

1 ← Atomic numer
H ← Symbol
Hydrogen ← Name
1.008 ← Atomic Weight

Legend: Halogen; Nonmetal; Alkali metal; Alkaline earth metal; Actinide; Transition metal; Other metal; Metalloid; Lanthanide; Noble gases

	1	2	3	4	5	6	7	8	9	10	11	12	13	14	15	16	17	18
1	1 H Hydrogen 1.008																	2 He Helium 4.0026
2	3 Li Lithium 6.94	4 Be Beryllium 9.01218											5 B Boron 10.81	6 C Carbon 12.011	7 N Nitrogen 14.007	8 O Oxygen 15.999	9 F Fluorine 18.9984	10 Ne Neon 20.1797
3	11 Na Sodium 22.98976	12 Mg Magnesium 24.305											13 Al Aluminium 26.9815	14 Si Silicon 28.085	15 P Phosphorus 30.97376	16 S Sulfur 32.06	17 Cl Chlorine 35.45	18 Ar Argon 39.948
4	19 K Potassium 39.0983	20 Ca Calcium 40.078	21 Sc Scandium 44.95591	22 Ti Titanium 47.867	23 V Vanadium 50.9415	24 Cr Chromium 51.9961	25 Mn Manganese 54.93804	26 Fe Iron 55.845	27 Co Cobalt 58.93319	28 Ni Nickel 58.6934	29 Cu Copper 63.546	30 Zn Zinc 65.38	31 Ga Gallium 69.723	32 Ge Germanium 72.63	33 As Arsenic 74.9216	34 Se Selenium 78.96	35 Br Bromine 79.904	36 Kr Krypton 83.798
5	37 Rb Rubidium 85.4678	38 Sr Strontium 87.62	39 Y Yttrium 88.9058	40 Zr Zirconium 91.224	41 Nb Niobium 92.9063	42 Mo Molybdenum 95.96	43 Tc Technetium (98)	44 Ru Ruthenium 101.07	45 Rh Rhodium 102.9055	46 Pd Palladium 106.42	47 Ag Silver 107.8682	48 Cd Cadmium 112.411	49 In Indium 114.818	50 Sn Tin 118.71	51 Sb Antimony 121.76	52 Te Tellurium 127.6	53 I Iodine 126.9044	54 Xe Xenon 131.293
6	55 Cs Caesium 132.9054	56 Ba Barium 137.327	56-71 Lanthanide	72 Hf Hafnium 178.49	73 Ta Tantalum 180.9478	74 W Tungsten 183.84	75 Re Rhenium 186.207	76 Os Osmium 190.23	77 Ir Iridium 192.217	78 Pt Platinum 195.084	79 Au Gold 196.9665	80 Hg Mercury 200.59	81 Tl Thallium 204.38	82 Pb Lead 207.2	83 Bi Bismuth 208.9804	84 Po Polonium (209)	85 At Astatine (210)	86 Rn Radon (222)
7	87 Fr Francium (223)	88 Ra Radium (226)	89-103 Actinide	104 Rf Rutherfordium (267)	105 Db Dubnium (268)	106 Sg Seaborgium (269)	107 Bh Bohrium (270)	108 Hs Hassium (269)	109 Mt Meitnerium (278)	110 Ds Darmstadtium (281)	111 Rg Roentgenium (281)	112 Cn Copernicium (285)	113 Uut Ununtrium (286)	114 Fl Flerovium (289)	115 Uuq Ununpentium (288)	116 Lv Livermorium (293)	117 Uus Ununseptium (294)	118 Uuo Ununoctium (294)

Lanthanide →	57 La Lanthanum 138.9054	58 Ce Cerium 140.116	59 Pr Praseodymium 140.9076	60 Nd Neodymium 144.242	61 Pm Promethium (145)	62 Sm Samarium 150.36	63 Eu Europium 151.964	64 Gd Gadolinium 157.25	65 Tb Terbium 158.9253	66 Dy Dysprosium 162.5	67 Ho Holmium 164.9303	68 Er Erbium 167.259	69 Tm Thulium 168.9342	70 Yb Ytterbium 173.054	71 Lu Lutetium 174.9668
Actinide →	89 Ac Actinium (227)	90 Th Thorium 232.038	91 Pa Protactinium 231.0358	92 U Uranium 238.0289	93 Np Neptunium (237)	94 Pu Plutonium (244)	95 Am Americium (243)	96 Cm Curium (247)	97 Bk Berkelium (247)	98 Cf Californium (251)	99 Es Einsteinium (252)	100 Fm Fermium (257)	101 Md Mendelevium (258)	102 No Nobelium (259)	103 Lr Roentgenium 281

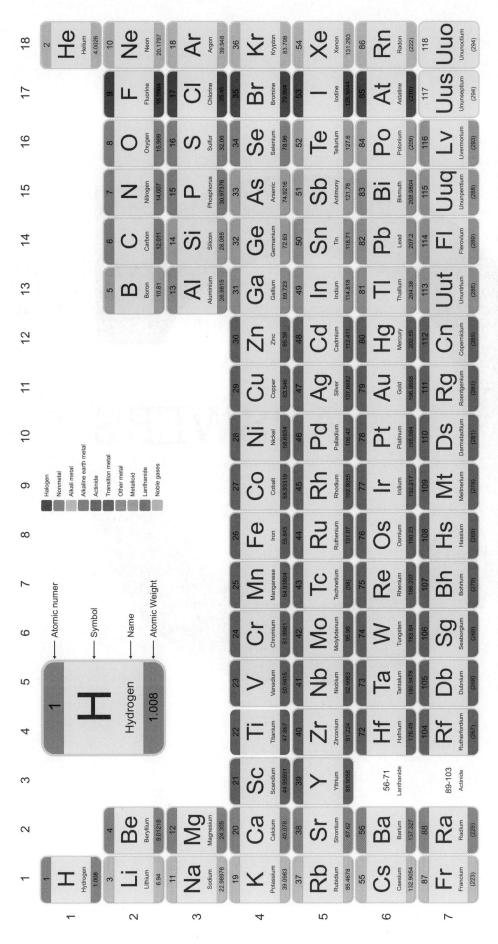

C ANSWERS

ISBN: 9780170352611

Chapter One

Pre-test

1 **a** 2, 8, 2
b 2, 8, 7
c 2, 8
d 2, 8, 8

2 It increases.

3 It decreases.

4 They go from forming positive to negative ions.

5 **a**

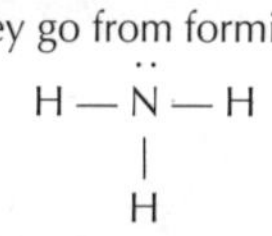

b

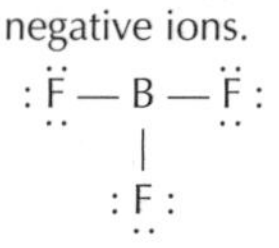

6 **a** ionic bonds
b metallic bonds
c intermolecular forces

7 The heat content of a substance and H is the letter that represents it.

8 n (mol) = 30 g/16 $gmol^{-1}$ = 1.875 mol
ΔH = 200 kJ/1.875 mol = 107 $kJmol^{-1}$ (3 s.f.)

1.1: Atom electron arrangement

a $1s^22s^22p^1$ or $[He]2s^22p^1$
b $1s^22s^22p^63s^2$ or $[Ne]3s^2$
c $1s^22s^22p^63s^23p^5$ or $[Ne]3s^23p^5$
d $1s^22s^22p^63s^23p^63d^14s^2$ or $[Ar]3d^14s^2$
e $1s^22s^22p^63s^23p^63d^64s^2$ or $[Ar]3d^64s^2$
f $1s^22s^22p^63s^23p^63d^{10}4s^2$ or $[Ar]3d^{10}4s^2$
g $1s^22s^22p^63s^23p^63d^54s^1$ or $[Ar]3d^54s^1$

1.2: Ion electron arrangement

a $1s^22s^22p^63s^23p^6$ or [Ar]
b $1s^22s^22p^6$ or [Ne]
c $1s^22s^22p^6$ or [Ne]
d $1s^22s^22p^63s^23p^63d^6$ or $[Ar]3d^6$
e $1s^22s^22p^63s^23p^63d^5$ or $[Ar]3d^5$
f $1s^22s^22p^63s^23p^63d^{10}$ or $[Ar]3d^{10}$
g $1s^22s^22p^63s^23p^63d^9$ or $[Ar]3d^9$

1.3: Orbital diagrams

a Hydrogen, H

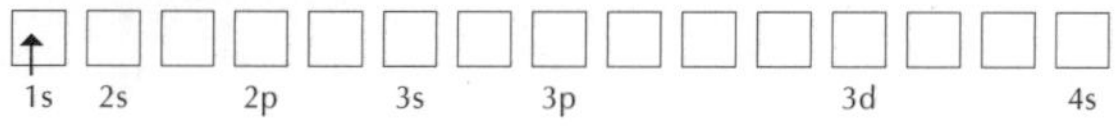

b Lithium, Li

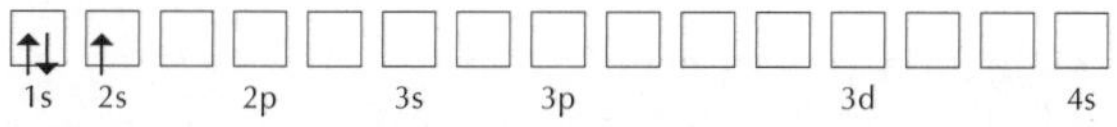

c Lithium ion, Li^+

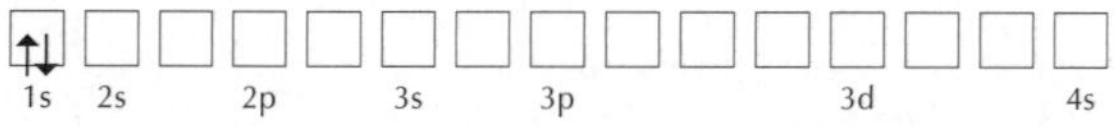

d Aluminium ion, Al^{3+}

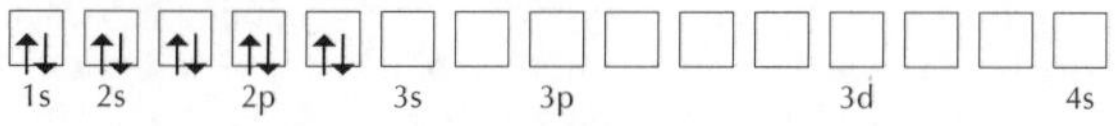

e Scandium, Sc

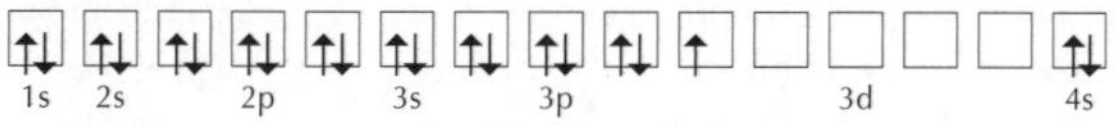

f Chromium, Cr (Don't forget it is an exception!)

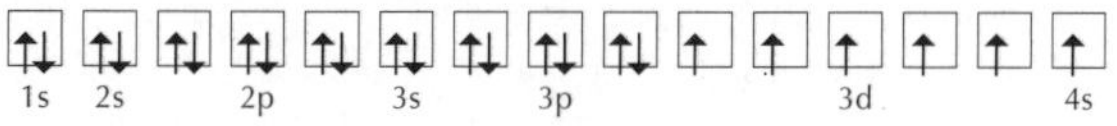

g Copper, Cu (Don't forget it is an exception!)

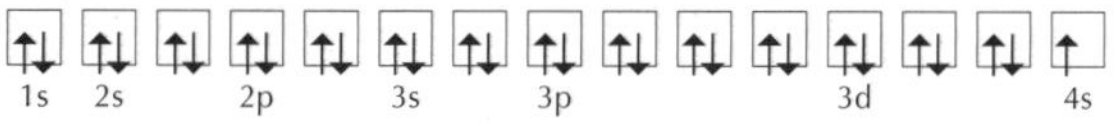

h Copper (II) ions, Cu^{2+}

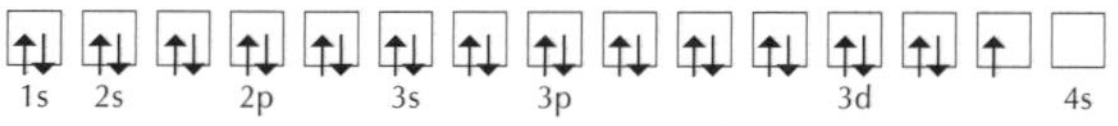

1.4: Electronegativity

a The ability for an atom to attract a bonding pair of electrons towards itself.

b **i** B — B has a higher electronegativity value than Li because B has a larger effective nuclear charge because it has more protons in its nucleus but its valence electrons are relatively no further from the nucleus than Li, which means that it has more ability to attract a bonding pair of electrons towards itself.

ii Li — Li has a higher electronegativity value than Na because it has one less layer of electrons than Na, so it has less electron shielding from the inner layers of electrons and so a greater effective nuclear charge, which means that it has more ability to attract a bonding pair of electrons towards itself.

iii F — F has a higher electronegativity value than Br because it has less layers of electrons than Br, so it has less electron shielding from the inner layers of electrons and so a greater effective nuclear charge, which means that it has more ability to attract a bonding pair of electrons towards itself.

1.5: Ionisation energy

a Ionisation energy is the amount of energy required to remove one mole of electrons from one mole of gaseous atoms.

b $Be_{(g)} \rightarrow Be^+_{(g)} + e^-$

c **i** B — B has a higher first ionisation energy than Li because the valence electrons are being added on to the same quantum level but there are more protons in the nucleus, therefore there is more effective nuclear charge and so more energy is required to remove a valence electron.

ii Li — Li has a higher first ionisation energy than Na because the valence electrons are closer to the nucleus as the atom contains one less quantum level of electrons, therefore there is more effective nuclear charge and less electron shielding and so more energy is required to remove a valence electron.

iii F — F has a higher first ionisation energy than Br because the valence electrons are closer to the nucleus as the atom contains less quantum levels of electrons, therefore there is more effective nuclear charge and less electron shielding and so more energy is required to remove a valence electron.

d Phosphorus has the electron arrangement $1s^22s^22p^63s^23p^3$ and so gets extra stability from having a half-filled p orbital than sulfur which has the electron arrangement $1s^22s^22p^63s^23p^4$ where it is not exactly half filled and so has increased repulsion between the valence electrons requiring slightly less energy to remove an electron from its atoms.

1.6: Atomic and ionic radii

a The distance between two bonded atomic nuclei divided by 2.

b **i** Li — Li has a larger atomic radius because it has less effective nuclear charge due to having less protons in its nucleus than B, and B has its valence electrons being no further from the nucleus, therefore it has more effective nuclear charge than Li. This leads to Li having a slightly larger atomic radius.

ii Na — Na has a larger atomic radius because the valence electrons are being shielded by the extra layer of inner electrons that it has, reducing its effective nuclear charge making it have more repulsion between its valence electrons and a larger size.

iii Br — Br has a larger atomic radius because the valence electrons are being shielded by the extra layers of inner electrons that it has, reducing its effective nuclear charge making it have more repulsion between its valence electrons and so a larger size.

c **i** Cl^- — Cl^- has a larger ionic radius because the valence electrons are being shielded by the extra layer of inner electrons that it has, reducing its effective nuclear charge making it have more repulsion between its valence electrons and so a larger size.

ii Cl^- — Cl^- has a larger ionic radius because it has more electrons to protons (unlike Na^+, which has more protons to electrons); this gives more electron-electron repulsion and so a larger size, plus Na^+ has one whole less layer of electrons.

Checkpoint 1: Electron arrangements and trends of the periodic table

1 a Ca $1s^22s^22p^63s^23p^64s^2$ or $[Ar]4s^2$
  b V $1s^22s^22p^63s^23p^63d^34s^2$ or $[Ar]3d^34s^2$
  c P $1s^22s^22p^63s^23p^3$ or $[Ne]3s^23p^3$
  d Si $1s^22s^22p^63s^23p^2$ or $[Ne]3s^23p^2$
  e Cu^+ $1s^22s^22p^63s^23p^63d^{10}$ or $[Ar]3d^{10}$
  f Be^{2+} $1s^2$ or [He]
  g S^{2-} $1s^22s^22p^63s^23p^6$ or [Ar]
  h Co^{2+} $1s^22s^22p^63s^23p^63d^7$ or $[Ar]3d^7$

2 O has one less layer of electrons shielding the effective nuclear charge from the valence electrons and any bonding pairs of electrons making it have a stronger pull on these electrons and so a higher electronegativity value.

3 Helium's valence electrons are closer to the nucleus than neon's, therefore there is less distance for the effective nuclear charge to work over, making it harder to pull one of the valence electrons off He than Ne and so a higher first ionisation energy.

4 The oxygen atom has the electron arrangement $1s^22s^22p^4$ whereas the oxide ion has the electron arrangement $1s^22s^22p^6$. These extra two valence electrons increase the repulsion force between the electrons making for a slightly larger outer orbital. There is also now less effective nuclear charge in the oxide ion as there are two less protons to electrons, which will also increase the size.

1.7: Expanded octet Lewis structures

a SF_6

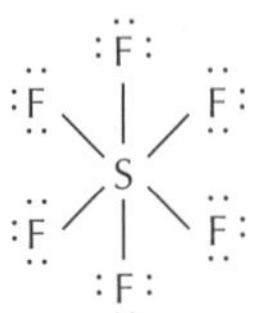

b XeF_4

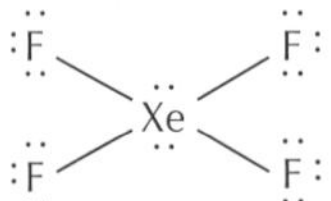

c ICl_3

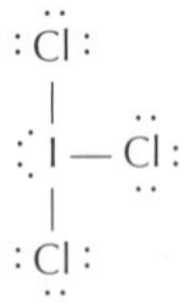

d SF_4

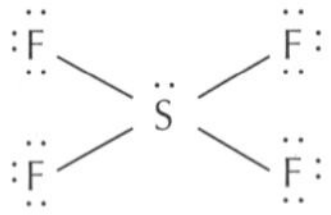

1.8: Polyatomic ion Lewis structures

a CO_3^{2-}

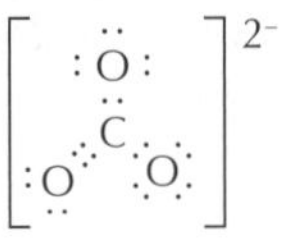

b NH_4^+

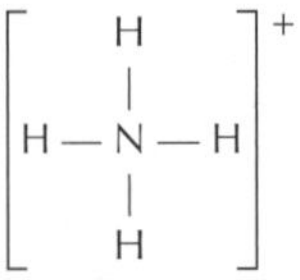

c SO_3^{2-}

d PCl_6^-

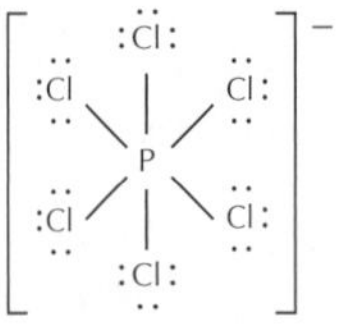

1.9: Shapes of molecules

Molecule	Lewis structure	Shape
CF_4	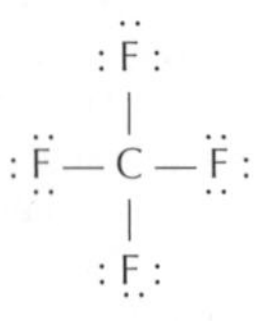	tetrahedral 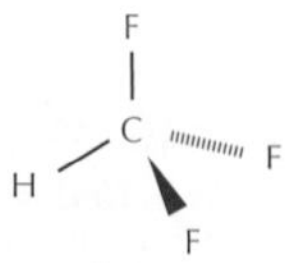
NF_3	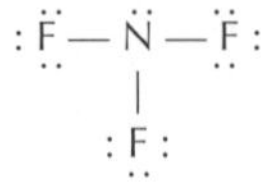	Trigonal pyramid 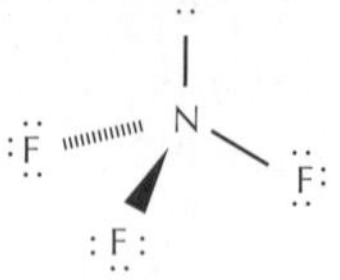
SF_3^+	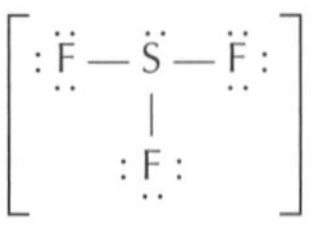	Trigonal pyramid 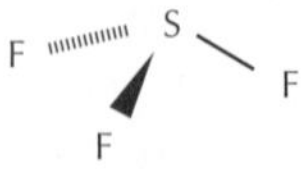
PCl_6^-	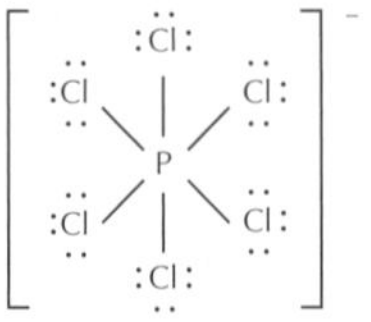	Octahedral 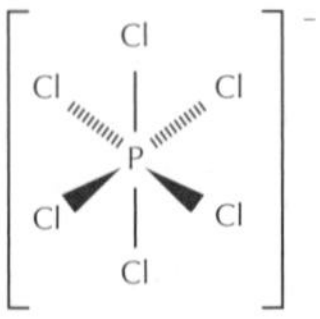
PCl_4^-	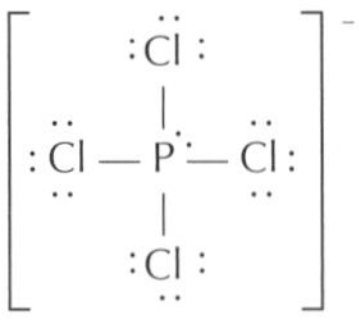	Seesaw or distorted tetrahedron 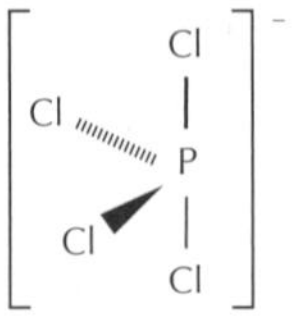
BrF_3	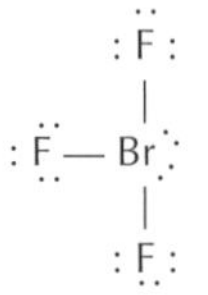	T-shaped 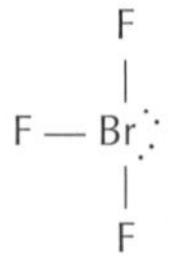
I_3^-	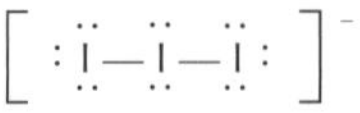	Linear
PH_3	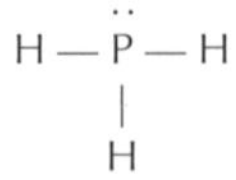	Trigonal pyramid 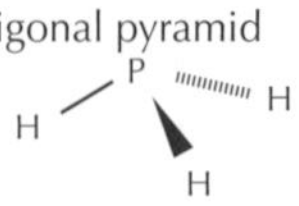

Investigation 1: Shape and polarity

1 ClF_3 has a T-shaped structure due to the two lone pairs and three bonding pairs around the central Cl atom. These lone pairs and bonding pairs repel each other so that they are as far apart as possible from one another giving bond angles of slightly less than 90° as the lone pairs repel more strongly than the bonded pairs. Due to this arrangement and number of bonding and non-bonding pairs, the molecule takes this shape.

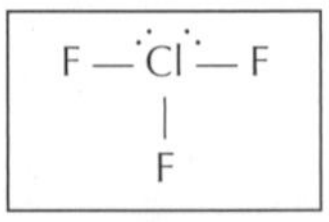

2 AsF_5 has a trigonal bipyramid shape due to the five bonding pairs of electrons about the central atom and the lack of lone pairs of electrons. These bonded electrons repel each other to be as far apart as possible giving bond angles of 90° and 120°. Due to the five bonded atoms and lack of lone pair, the molecule has this shape.

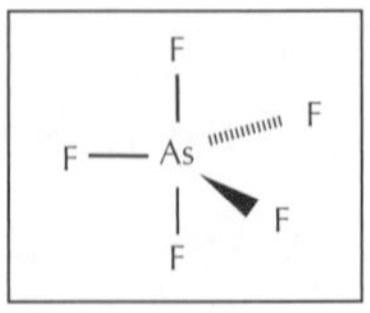

3 NF_3 has a trigonal pyramid shape. The molecule contains polar bonds due to the electronegativity difference between the N and the F atom, with the F being more electronegative than the N atom and so being slightly negative. As the molecule is asymmetrical overall, the dipoles do not cancel and so the molecule is polar overall.

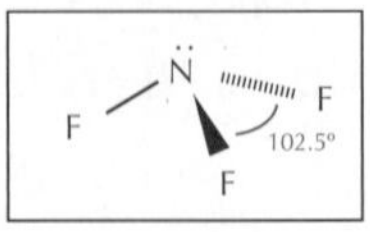

ISBN: 9780170352611

4 IF_5 has a square pyramidal shape. The molecule contains polar bonds due to the electronegativity difference between the I and the F atoms with the F being more electronegative than the I making it slightly negative. As the molecule is overall asymmetrical and the molecule contains polar bonds, it is overall polar.

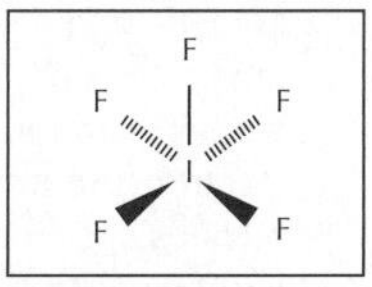

1.10: Forces of attraction

a **i** hydrogen bonds, covalent bonds, permanent dipole-dipole bonds
ii covalent and instantaneous dipole-induced dipole
iii covalent and instantaneous dipole-induced dipole
iv ionic
v metallic

b H_2O has the much stronger hydrogen bonds between its molecules, which take a larger amount of energy to break rather than just the permanent dipole-dipole bonds that are present in between the molecules of H_2S.

c Ammonia has the much stronger hydrogen bonds between its molecules, which take a larger amount of energy to break than the permanent dipole-dipole bonds between the phosphane molecules.

d All of the molecules contain instantaneous dipole-induced dipole bonds, which are the bonds that are broken when the substance melts. As the length of chain gets longer, the melting point increases because there are more opportunities along the length of the molecule for instantaneous dipole-induced dipole bonds to occur, requiring more energy for the substance to melt.

Experiment 1: The effect of the hydrogen bond

Because water forms strong hydrogen bonds with the ethanol allowing the molecules to get closer together, hence the decrease in volume, and as bond making is exothermic there is also a rise in temperature, as energy is required for bonds to form. As cyclohexane is non-polar there are interactions formed between it and the water molecules.

Checkpoint 2: Lewis structures, shapes and types of bonds

1 a

PCl_5	SF_4	XeF_4
Lewis structure 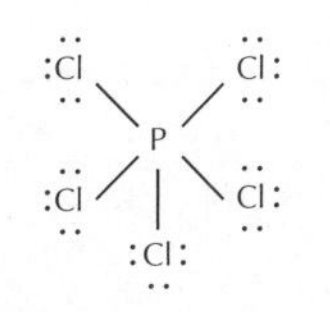	Lewis structure 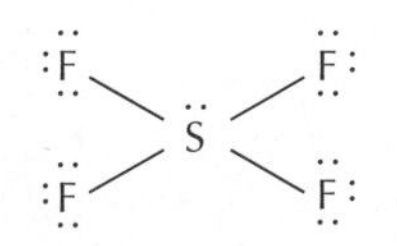	Lewis structure 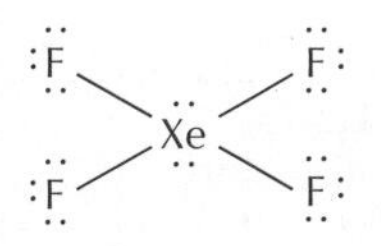
Shape Trigonal bipyramidal	**Shape** Seesaw or distorted tetrahedron	**Shape** Square planar

b SF_4 has one lone pair and four bonding pairs which repel each other to be as far apart as possible giving bond angles of slightly less than 90° and 120°, with the lone pair repeling more strongly than the bonded pairs of electrons. XeF_4 has six regions of negative charge about its central atom (one more than SF_4) with two lone pairs and four bonded pairs of electrons. These regions repel each other to be as far apart as possible giving bond angles of 90°, with the lone pairs repelling more strongly than the bonding pairs of electrons.
Because these two structures have different numbers of regions of negative charge about their central atoms, they have different shapes.

2 Due to the hydrogen bonds that are present between the molecules of H_2O and HF they have significantly higher boiling points as these bonds require more energy to break when the substance is boiled than the permanent dipole-dipole bonds which are in all the other substances. As the atoms in the molecules get larger, like with H_2Te and HI, there is more surface area for bonds to occur between the molecules and so they have higher boiling points than the molecules formed with atoms higher up their groups of the periodic table.

1.11: Specific heat capacity

a $q = mc\Delta T = 50 \times 4.18 \times (90 - 25) = 13{,}585\ J = 13{,}600\ J$ or 13.6 kJ (3 s.f.)

b $q = mc\Delta T$
$2500 = 30.2 \times 4.18 \times \Delta T$
$\Delta T = 2500/126.236 = 19.8°C$ (3 s.f.)

c $q = mc\Delta T = 930 \times 4.18 \times (52 - 18) = 132{,}171.6\ J = 132{,}000\ J$ or 132 kJ (3 s.f.)

d **i** $q = mc\Delta T = 250 \times 4.18 \times (30 - 20) = 10{,}450\ J$ or 10,500 J or 10.5 kJ (3 s.f.)
Note: We ignore the mass of CaO here because it does not relate to the specific heat capacity of water.
ii $n(CaO) = mass/M_r = 6.0/56.0 = 0.1071428571$ mol (Don't round here)
$\Delta_r H = q/moles = 10.45\ kJ/ 0.1071428571\ mol = 97.5\ kJmol^{-1}$
Note: The q value needs to be in kJ and unrounded for this part of the calculation.

e $q = mc\Delta T$
$1621 = 38 \times 4.18 \times (34 - T_f)$
$34 - T_f = 1621/(38 \times 4.18)$
$T_f = 34 - 10.205... = 23.8°C$ (3 s.f.)

f **i** $q = mc\Delta T = 250 \times 4.18 \times (25.5 - 21.0) = 4702.5\ J$ or 4.7025 kJ
ii $n(Mg(OH)_2) = mass/M_r = 7.00/40.3 = 0.17369...$ mol
$\Delta_r H = q/moles = 4.7025/0.17369... = 27.1\ kJmol^{-1}$ (3 s.f.)

1.12: Enthalpy of formation

a $H_{2(g)} + ½O_{2(g)} \longrightarrow H_2O_{(l)}$
b $C_{(s)} + O_{2(g)} \longrightarrow CO_{2(g)}$
c $5C_{(s)} + 6H_{2(g)} \longrightarrow C_5H_{12(l)}$
d $Mg_{(s)} + ½O_{2(g)} \longrightarrow MgO_{(s)}$
e $Ca_{(s)} + C_{(s)} + 1½O_{2(g)} \longrightarrow CaCO_{3(s)}$

1.13: Enthalpy of combustion

a $H_{2(g)} + ½O_{2(g)} \longrightarrow H_2O_{(g)}$
b $CH_{4(g)} + 2O_{2(g)} \longrightarrow CO_{2(g)} + 2H_2O_{(g)}$
c $C_5H_{12(l)} + 8O_{2(g)} \longrightarrow 5CO_{2(g)} + 6H_2O_{(g)}$
d $C_{12}H_{26(s)} + 18½O_{2(g)} \longrightarrow 12CO_{2(g)} + 13H_2O_{(g)}$
e $C_3H_7OH_{(l)} + 4½O_{2(g)} \longrightarrow 3CO_{2(g)} + 4H_2O_{(g)}$

1.14: Enthalpies of change of state

a $H_2O_{(l)} \longrightarrow H_2O_{(g)}$
b $H_2O_{(s)} \longrightarrow H_2O_{(g)}$
c $CO_{(l)} \longrightarrow CO_{(g)}$
d $CO_{2(s)} \longrightarrow CO_{2(l)}$
e $C_4H_{10(s)} \longrightarrow C_4H_{10(g)}$

1.15: Entropy

a Increases
b Increases
c Decreases
d **i** Increases
ii There are more gaseous particles present.
e **i** Increases
ii There are more gaseous particles present and no solid particles present.

1.16: Hess's Law

a $\Delta H = -1410 + 1560 - 286 = -136\ kJmol^{-1}$

b $\Delta H = 2 \times -181 + 2 \times 91.8 + 3 \times -484 = -1630\ kJmol^{-1}$ (3 s.f.)

c $\Delta_f H°(CH_3COOH) = ?$ $2C_{(s)} + 2H_{2(g)} + O_{2(g)} \longrightarrow CH_3COOH_{(l)}$
$\Delta_f H(CO_2) = -395\ kJmol^{-1}$ $C_{(s)} + O_{2(g)} \longrightarrow CO_{2(g)}$
$\Delta_f H(H_2O) = -286\ kJmol^{-1}$ $H_{2(g)} + ½O_{2(g)} \longrightarrow H_2O_{(l)}$
$\Delta_f H°(CH_3COOH) = 875 + 2 \times -395 + 2 \times -286 = -487\ kJmol^{-1}$

d $C_{(s)} + 2H_{2(g)} \longrightarrow CH_{4(g)}$ $\Delta_f H°(CH_{4(g)}) = -74.9\ kJmol^{-1}$
$\Delta H = ½ \times 91.8 + 74.9 + ½ \times 270.3 = 256\ kJmol^{-1}$ (3 s.f.)

e $\Delta H = -1050 + 6 \times -74.8 + 3 \times -1850 + 2 \times 323 = -6400\ kJmol^{-1}$ (3 s.f.)

Experiment 2: Hess's Law

$M(MgSO_4.7H_2O) = 246.4\ gmol^{-1}$
$M(MgSO_4) = 120.4\ gmol^{-1}$

Questions

1 The two experimental equations are:
$MgSO_{4\,(s)} \longrightarrow Mg^{2+}_{(aq)} + SO^{2-}_{4\,(aq)}$
$q = mc\Delta T = 7.5 \times 4.18 \times \Delta T$ (from the experiment) — remember to convert the answer into kJ by dividing by 1000.
$n(MgSO_4)$ = mass/molar mass = 7.5/120.4 = 0.06229… mol
$\Delta H = q/n$
$MgSO_4.7H_2O_{(s)} \longrightarrow Mg^{2+}_{(aq)} + SO^{2-}_{4\,(aq)}$
$q = mc\Delta T = 15.5 \times 4.18 \times \Delta T$ (from the experiment) — remember to convert the answer into kJ by dividing by 1000.
$n(MgSO_4.7H_2O)$ = mass/molar mass = 15.5/246.4 = 0.0629… mol
$\Delta H = q/n$

2 ΔH = -84 -16 = -100 $kJmol^{-1}$

3 They may be different due to different conditions in the classroom from the ones that are done under standard conditions. There might be experimental error.

Checkpoint 3: Specific heat capacity and Hess's Law

1 $q = mc\Delta T = (55 + 45) \times 4.18 \times (25.5 - 21.0)$ = 1881 J = 1880 J (3 s.f.)

2 n(ice) mass/M_r = 75.0/18 = 4.166… mol
$q = \Delta H \times n = 6.02 \times 4.1666\ldots$ = 25.1 kJ (3 s.f.)

3 ΔH = -297 + -900 + -242 = -1439 $kJmol^{-1}$

4 ΔH = 75 +-393 + 2 x -242 = -802 $kJmol^{-1}$

Exam-type questions

Question one

a Cu $1s^22s^22p^63s^23p^63d^94s^1$ or $[Ar]3d^94s^1$
Fe^{3+} $1s^22s^22p^63s^23p^63d^5$ or $[Ar]3d^5$

b As you go down the group the size of the atom increases as there is another energy level each time with relatively no more nuclear charge and so due to these extra electrons shielding the nuclear charge the atoms get larger.

c The first ionisation energy of bromine is lower than chlorine due to the extra energy level of electrons that shield the nuclear charge from the valence electrons and so make the force of attraction weaker. This means it takes less energy to remove a valence electron from bromine.

d There are two reasons why the second ionisation energy is greater, the first is there are now 11 protons to 10 electrons meaning there is more effective nuclear charge acting over a shorter distance. It is a shorter distance because the outer energy level of electrons has now been completely lost making the valence electrons that much closer to the nucleus.

Question two

a Cl $1s^22s^22p^63s^23p^5$ or $[Ne]3s^23p^5$
Cr^{3+} $1s^22s^22p^63s^23p^63d^3$ or $[Ar]3d^3$

b The atomic radii decrease as you go across the period as the valence electrons are being added on to the same energy level and so are no further from the nucleus but there is an increase in nuclear charge as there are more protons found in the nucleus.

c As you go across period 3 of the periodic table, electronegativity increases as the valence electrons are no further from the nucleus as they are being added on to the same energy level, but there is a corresponding increase in nuclear charge as there are more protons being added. This means that there is more effective nuclear charge acting over the same distance, which means there will be more effective nuclear charge acting on any electrons in a bond.

d The fluorine atom is much smaller than the fluoride ion as there is more effective nuclear charge, as there are the same amount of protons to electrons. Once it becomes fluoride there is one more electron, which increases the repulsion between the valence electrons pushing them further from the nuclear charge.

Question three

a Aluminium has a greater atomic radius than phosphorus because P has more effective nuclear charge due to the valence electrons being no further from the nucleus as its electrons are in the same energy level, while also having more protons in its nucleus so there will be more effective nuclear charge acting on the valence electrons and so a smaller size. Once aluminium becomes an ion it loses a whole energy level and it loses three electrons, so it now has a valence layer that is closer to the nucleus with more protons acting on it. When P becomes an ion it gains three electrons, increasing the electron repulsion in the valence layer and making it have a slightly larger distance from the nucleus. There is also less effective nuclear charge because there are less protons to electrons, making the P^{3-} ion much larger than its atom.

b As you go across the period from B to N, there is an increase in first ionisation energy because there is an increase in effective nuclear charge as valence electrons are being added to the same layer and so are no further from the nucleus, but there are more protons in the nucleus increasing the effective nuclear charge and so it takes more energy to remove an electron.

Question four

a

	SF_3^-	SO_4^{2-}	SO_3^{2-}
Lewis diagrams	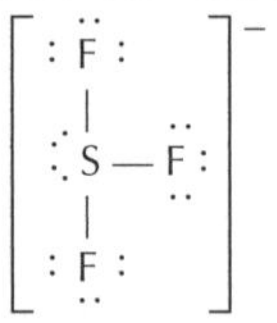	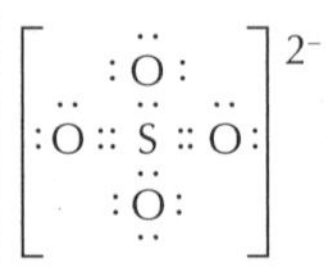	
Shape	T-shaped	Tetrahedral	Trigonal pyramid

b SF_3^- has five regions of negative charge about its central atom, three bonding pairs and two lone pairs of electrons; these electrons repel each other to be as far apart as possible giving a T shape.
SO_4^{2-} has also got four regions of negative charge about the central atom, two single bonds and two double bonds; these regions repel each other to be as far apart as possible giving a tetrahedral shape.
SO_3^{2-} also has four regions of negative charge about the central atom with one lone pair and three bonding pairs of electrons which repel each other to be as far apart as possible giving a trigonal pyramid shape.

Question five

a

	PCl_5	ICl_5
Lewis diagrams	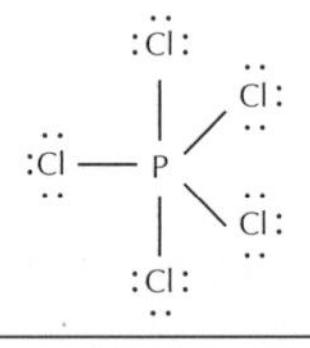	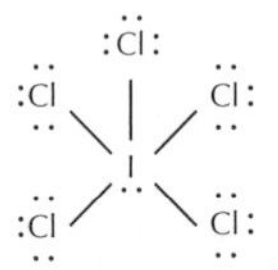
Shape	Trigonal bypyramid	Square pyramid
Polar or non-polar	Non-polar	Polar

b PCl_5 has five regions of negative charge about the central atom; all of them are bonding pairs which repel each other to be as far apart as possible giving a trigonal bypyramid shape. PCl_5 contains polar bonds due to the electronegativity difference between the P and the Cl, where the Cl is slightly negative as it is the more electronegative atom. However as the molecule is symmetrical overall, the bond dipoles cancel leaving the molecule non-polar overall.
In contrast, ICl_5 has six regions of negative charge about the central atom, five bonding pairs and one lone pair which repel each other to be as far apart as possible leaving a square pyramidal shape. It also contains polar bonds due to the electronegativity difference between the I and the Cl with the Cl being more electronegative than I so it becomes slightly negative. As the molecule is not symmetrical though, the dipoles do not cancel allowing the molecule to be polar overall.

ISBN: 9780170352611

Question six

a

	SF_4	XeF_4
Lewis diagrams	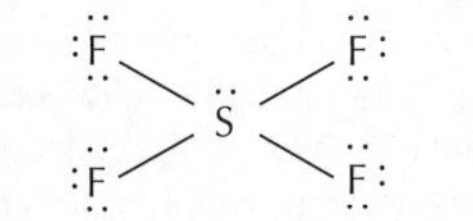	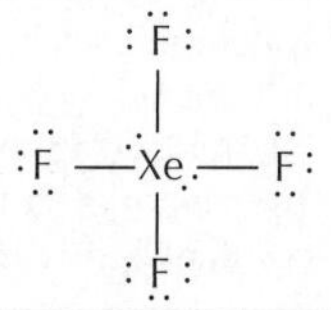
Shape and shape diagram	Seesaw/distorted tetrahedron 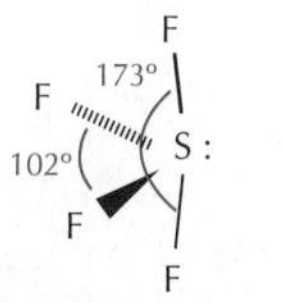	Square planar 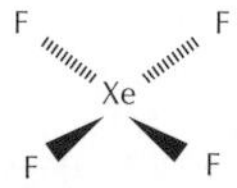
Polar or non-polar	Polar	Non-polar

b SF_4 has five regions of negative charge about the central S atom, with four bonding pairs and one lone pair of electrons which repel each other so that they are as far apart as possible giving a seesaw shape. SF_4 contains polar bonds due to the difference in electronegativity between the S and the F atom, with the F being more electronegative than the S and so it is slightly negative. As the molecule is asymmetrical in shape, these bond dipoles do not cancel making the molecule polar overall. XeF_4 has six regions of negative charge about the central Xe atom, with four bonding pairs and two lone pairs of electrons which repel each other to be as far apart as possible giving a square planar shape. XeF_4 contains polar bonds due to the electronegativity difference between the Xe and the F atoms, because F is more electronegative it will be slightly negative. Due to the molecule being symmetrical though, these bond dipoles cancel leaving the molecule non-polar overall.

Question seven

a The $\Delta_{sub}H^o$ is the amount of energy required to convert one mole of a solid into gaseous state under standard conditions of room temperature and pressure.

b Room temperature, 25°C and pressure 1 atmosphere.

c $\Delta H = 1368 + 2 \times -393 + 3 \times -242 = -144$ kJmol^{-1}

Question eight

a $\Delta_f H^o$ means the amount of energy to make one mole of a substance from its elements in their standard states.

b $C_{(s)} + O_{2\,(g)} \longrightarrow CO_{2\,(g)}$

$H_{2\,(g)} + \frac{1}{2}O_{2\,(g)} \longrightarrow H_2O_{(g)}$

c $\Delta_r H^o = 126 + 4 \times -394 + 5 \times -242 = -2660$ kJmol^{-1}

Question nine

a i $q = mc\Delta T$

$2760 = 110 \times 4.18 \times (T_f - 21)$

$T_f = 28.6$°C

ii $n(HCl) = \text{mass}/M_r = 55/36.5 = 1.50\ldots$ mol

$\Delta_r H = q/\text{moles}$

$\Delta_r H = 2.76/1.50\ldots = 1.83$ kJmol^{-1}

b $\Delta H = 3 \times 92 + 174 + 52 + 2 \times -286 = -70$ kJmol^{-1}

Question ten

a i $CH_{4\,(l)} \longrightarrow CH_{4\,(g)}$

ii Ammonia contains hydrogen bonds between the N and H on corresponding molecules; these forces are strong and require larger amounts of energy to break than the instantaneous dipole-induced dipole bonds between the methane molecules, despite the fact that both molecules have similar relative masses.

iii This is because it takes more energy for all the hydrogen bonds between the ammonia molecules to break in order for ammonia to turn into a gas, than it does to break all of the instantaneous dipole-induced dipole bonds between the methane molecules.

iv The enthalpy is positive because it is an endothermic reaction and so requires energy in order for the reaction to occur; this is because it takes energy in order to break the instantaneous dipole-dipole bonds. The entropy will increase as this reaction proceeds because there is an increase in the amount of arrangements in space for gaseous molecules than liquid ones, so there is an increase in randomness for the particles.

Chapter Two

Pre-test

1

	Name	Condensed structural formula
a	2,2-dimethylbutane	$CH_3C(CH_3)_2CH_2CH_3$
b	2-chloropropane	$CH_3CH(Cl)CH_3$
c	Ethanoic acid	CH_3COOH
d	2-methylpropan-1-ol	$CH_3CH(CH_3)CH_2OH$

	Molecular and empirical formula	Functional group
a	C_6H_{14}, C_3H_7	Alkane
b	C_3H_7Cl	Haloalkane
c	$C_2H_4O_2$, CH_2O	Carboxylic acid
d	$C_4H_{10}O$	Alcohol

2 a dilute H_2SO_4 and heat

b Structure: $CH_3CH_2CH_2Cl$

Name: 1-chloropropane (minor product only drawn)

c NH_3 warm

d Name: Propan-1-amine or 1-aminopropane

3 $CH_3CH_2CH_2CH_2CH_3$ pentane

$CH_3CH(CH_3)CH_2CH_3$ 2-methylbutane

$C(CH_3)_4$ dimethylpropane

2.1: Isomers

1

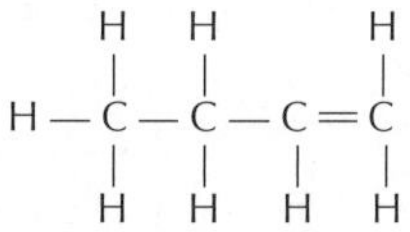

but-1-ene

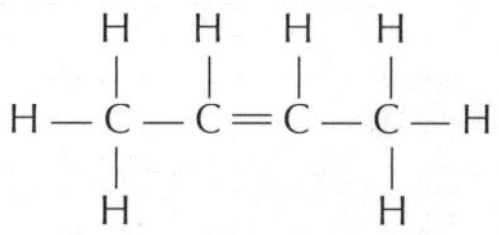

cis-but-2-ene

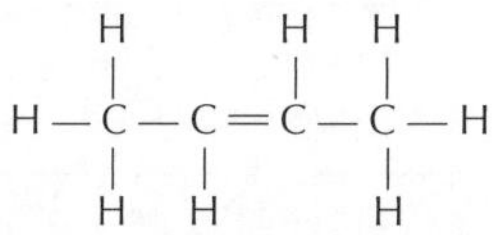

trans-but-2-ene

A structural isomer is a structure with the same molecular formula, in this case C_4H_{10}, but a different arrangement of the atoms in space; here the double bond is moved.

A geometric isomer is a special case of structural isomer that forms because there is no free rotation about a double bond and there are two different groups attached to each carbon in the double bond. In this case but-2-ene has a H and a CH_3 attached to each carbon in the double bond and so it can form geometric isomers, whereas but-1-ene has one carbon connected to two hydrogen atoms and so it cannot form geometric isomers.

2 a i

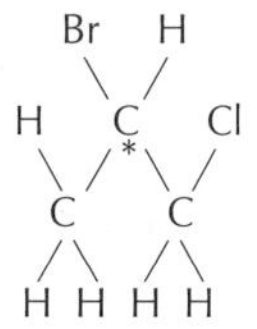

ISBN: 9780170352611

b

c i Compound can form optical isomers because it has a chiral carbon, that is a carbon with four different groups attached to it. The two enantiomers can be separated by shining a plane of polarised light through samples of each; the light will rotate in different directions depending on which isomer is present.

3 a ii

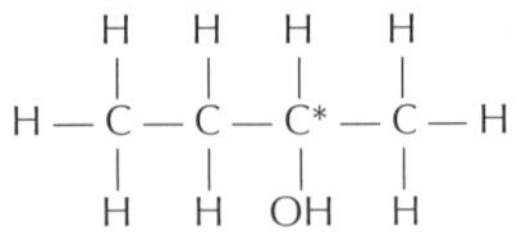

b

c ii Compound can form optical isomers because it has a chiral carbon, that is a carbon with four different groups attached to it. The two enantiomers can be separated by shining a plane of polarised light through samples of each; the light will rotate in different directions depending on which isomer is present.

2.2: Hydrocarbons

1 a propane
b 2,2-dimethylpropane
c 2-methylbutane
d $CH_3CH(CH_3)CH_2CH(CH_3)CH_3$
e $CH_3CH(CH_3)CH_2(CH_2)_2CH_3$
f propene
g 2-methylbut-2-ene
h 3-methylpent-2-ene

2 a $CH_2CH_2 + Br_2 \longrightarrow CH_2BrCH_2Br$
b CH_3CHCH_2 + dil. $H_2SO_4 \xrightarrow{heat} CH_3CH(OH)CH_3$ (major) + $CH_3CH_2CH_2OH$ (minor)
c $CH_3CH_3 + Br_2 \xrightarrow{UV} CH_3CH_2Br$
d $nCH_2CHCH_2CH_3 \longrightarrow [-CH_2 - CH(CH_2CH_3) - CH_2 - CH(CH_2CH_3) -]_n$
e $CH_3CH{=}CHCH_3 + KMnO_4/H^+ \longrightarrow CH_3CH(OH)CH(OH)CH_3$

2.3: Alcohols

1

Name	Structure
2-methylpropan-2-ol	$CH_3C(OH)(CH_3)_2$
3-methylbutan-1-ol	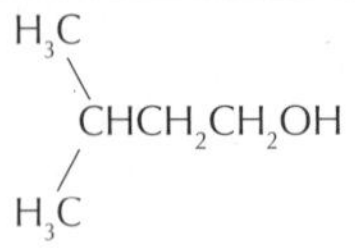
3-chlorobutan-1-ol	$CH_3CH(Cl)CH_2CH_2OH$

2 a $CH_3CH_2CH_2OH$ + cold $H^+/K_2Cr_2O_7 \longrightarrow CH_3CH_2COH$
b $CH_3CH_2CH(OH)CH_3$ + reflux/$H^+/K_2Cr_2O_7 \longrightarrow CH_3CH_2C(O)CH_3$
c $CH_3CH_2CH_2CH(OH)CH_3$ + 170ºC/c. $H_2SO_4 \longrightarrow CH_3CH_2CH_2CH{=}CH_2$ (minor) + $CH_3CH_2CH{=}CHCH_3$ (major)
d $CH_3CH_2OH + SOCl_2 \longrightarrow CH_3CH_2Cl$
e CH_3CH_2OH + reflux/$H^+/K_2Cr_2O_7 \longrightarrow CH_3COOH$
f CH_3CH_2OH + cold $H^+/K_2Cr_2O_7 \longrightarrow CH_3COH$
g $CH_3CH(OH)CH_3$ + reflux/$H^+/K_2Cr_2O_7 \longrightarrow CH_3C(O)CH_3$
h $CH_3CH_2OH + Al_2O_3 \longrightarrow CH_2{=}CH_2$
i $CH_3CH_3CH_2CH_2OH + SOCl_2 \longrightarrow CH_3CH_3CH_2CH_2Cl$
j $CH_3CH_3CH_2CH_2OH$ + reflux/$H^+/K_2Cr_2O_7 \longrightarrow CH_3CH_3CH_2COOH$

Experiment 1: Properties of alcohols

Questions

1 All three alcohols are soluble in water.

2 They all contain hydrogen intermolecular bonds as does the water molecules. So the attraction between the water and the alcohol molecules is strong enough to overcome the intermolecular bonds between the alcohol molecules.
The intra-molecular bonds are a mixture of polar and non-polar covalent bonds and these do not break when the alcohols dissolve.
This is because the forces of attraction between the alcohol and the water molecules is stronger than the attraction between the alcohol molecules so the alcohols then dissolve in water.

3 $CH_3CH_2OH + H^+/K_2Cr_2O_7 \longrightarrow CH_3COH$; if heated this molecule will go straight to the carboxylic acid CH_3COOH
$CH_3CH_2CH_2CH_2OH + H^+/K_2Cr_2O_7 \longrightarrow CH_3CH_2CH_2COH \longrightarrow CH_3CH_2CH_2COOH$
There will be no reaction with the 2-methylpropan-2-ol as it is a tertiary alcohol.

4 As there is a reaction with potassium dichromate with both ethanol and butan-1-ol, these are the primary alcohols.

5 2-methylpropan-2-ol is the tertiary alcohol because it does not react with the acidified potassium dichromate.

Checkpoint 1: Isomers, hydrocarbons and alcohols

1

Name	Structure
2-methylpropan-1-ol	$CH_3CH(CH_3)CH_2OH$
2-methylbut-2-ene	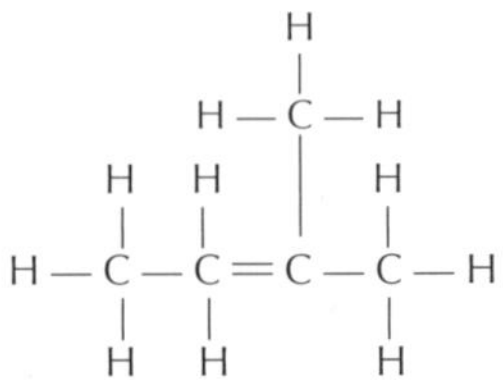
2,2-dimethylhexane	$CH_3C(CH_3)_2CH_2CH_2CH_2CH_3$
3-methylhex-1-ene	$CH_2{=}CHCH(CH_3)CH_2CH_2CH_3$

2

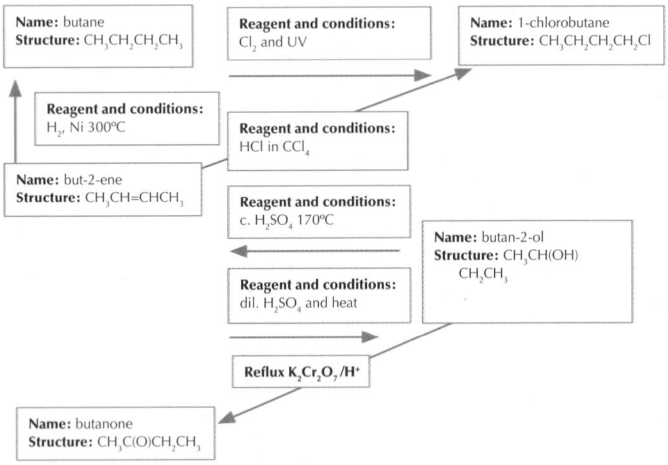

3 a i

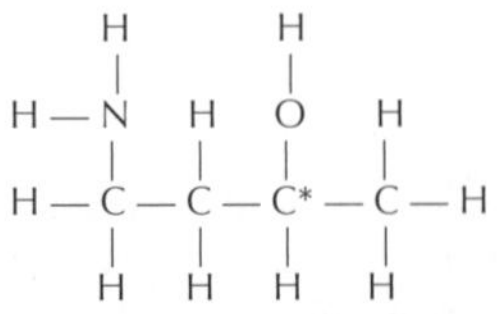

This molecule does because it contains an asymmetric carbon (shown by the *) that is a carbon with four different groups attached to it.

b

ISBN: 9780170352611

c The two enantiomers could be separated out by shining a plane of polarised light at samples of each; each sample will rotate the plane of polarised light in different directions.

4 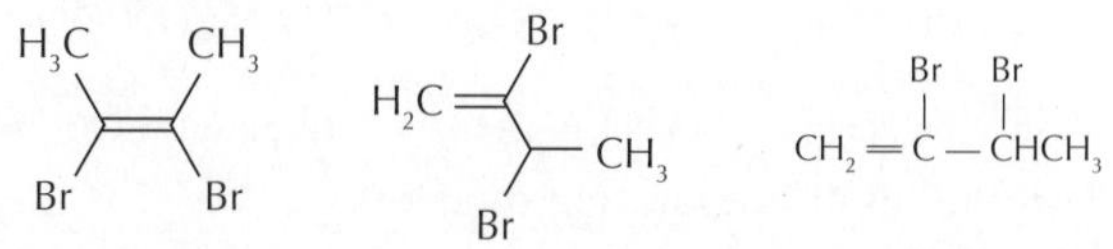

cis-2,3-dibromobut-2-ene *trans*-2,3-dibromobut-2-ene 2,3-dibromobut-1-ene

These two molecules are structural isomers of each other because they have the same molecular formula $C_4Br_2H_6$ but a different arrangement in space. 2,3-dibromobut-1-ene cannot form geometric isomers because it does not have two carbons in the double bond, which have two different groups attached to each.

2,3-dibromobut-2-ene can form geometric isomers because it has two different groups of atoms or atoms attached to each carbon in its double bond, and because there is no free rotation about a double bond, two different structures can be formed.

2.4: Haloalkanes

1

Name	Structure
2-bromo-2-methylpropane	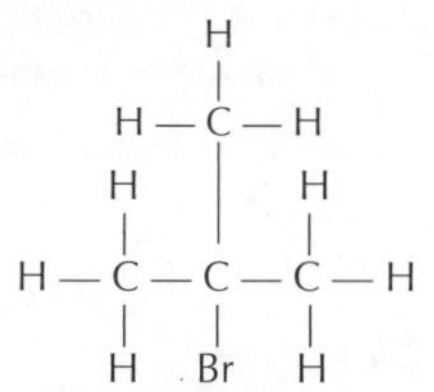
2-bromo-3-chlorobutane	$CH_3CH(Br)CH(Cl)CH_3$
2-chloro-3-methylbutane	$CH_3CH(Cl)CH(CH_3)_2$

2

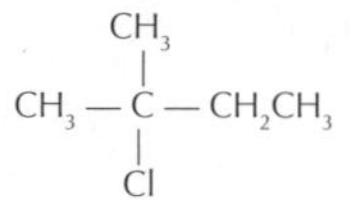	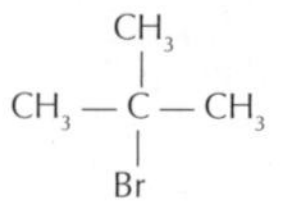	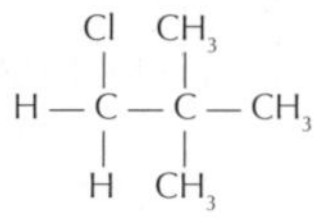
Tertiary	Tertiary	Primary

3 a $CH_3CH_2CH_3 + Cl_2 \xrightarrow{UV} CH_3CH_2CH_2Cl$ (Note: Any of the carbons could be substituted.)

b $CH_3CH{=}CH_2 + HCl\ (CCl_4) \longrightarrow \underset{major}{CH_3CH(Cl)CH_3} + \underset{minor}{CH_3CH_2CH_2Cl}$

c $CH_3CH_2CH_2Br + NH_3$ (warm) $\longrightarrow CH_3CH_2CH_2NH_2$

d $CH_3CH(CH_3)CH_2Br + KOH_{(aq)} \xrightarrow{reflux} CH_3CH(CH_3)CH_2H$

e $CH_3C(Br)(CH_3)_2 + KOH_{(alc)} \xrightarrow{reflux} CH_2{=}C(CH_3)_2$

2.5: Amines

1

Name	Structure
2-methylbutan-1-amine	$CH_3CH_2CH(CH_3)CH_2NH_2$
methanamine	CH_3NH_2
2-methylpropan-1-amine	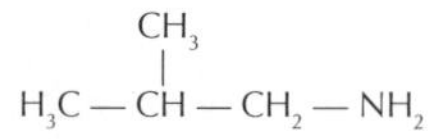

2

Structure	Classification
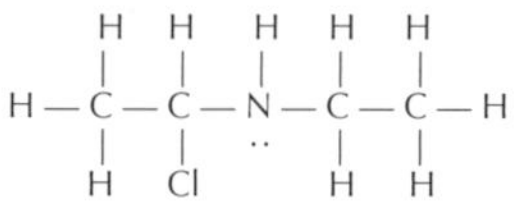	Secondary
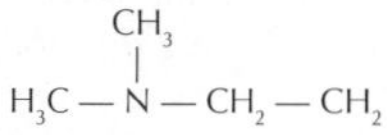	Tertiary
$CH_3CH_2CH(NH_2)CH_3$	Secondary
CH_3NH_2	Primary

3 a $CH_3CH_2NH_2 + HNO_3 \longrightarrow CH_3CH_2NH_3^+NO_3^-$

b $CH_3CH_2NH_2 + H_2O \rightleftharpoons CH_3CH_2NH_3^+ + OH^-$

c $CH_3CH_2Br + NH_3$ (warm) $\longrightarrow CH_3CH_2NH_2$

Experiment 2: Amines

Questions

1 $CH_3CH_2CH_2NH_2 + H_2O \rightleftharpoons CH_3CH_2CH_2NH_3^+ + OH^-$

2 The shorter chain will be more soluble because it has less instantaneous dipole-induced dipole forces acting so that the hydrogen bond between the NH of neighbouring molecules has more impact therefore forming strong forces of attraction between the water molecules.

2.6: Aldehydes and ketones

1

Name	Structure
Pentan-2-one	$CH_3C(O)CH_2CH_2CH_3$
Pentanal	$CH_3CH_2CH_2CH_2COH$
Butanone	$CH_3CH_2C(O)CH_3$
propanal	CH_3CH_2COH

2 a CH_3CH_2COH + Benedict's solution $\longrightarrow CH_3CH_2COOH$

b $CH_3CH(CH_3)COH$ + Fehling's solution $\longrightarrow$ $CH_3CH(CH_3)COOH$

c CH_3COH + Tollens' reagent $\longrightarrow CH_3COOH$

d CH_3CH_2OH + cold $K_2Cr_2O_7/H^+ \longrightarrow CH_3COH$

e $CH_3CH(OH)CH_3$ + reflux $K_2Cr_2O_7/H^+ \longrightarrow CH_3C(O)CH_3$

3 Add small amounts of butanone and butanal to Benedict's solution and heat gently over a Bunsen flame; the butanal sample should turn from a blue due to the Cu^{2+} ions in the Benedict's to a brick-red precipitate of Cu_2O as the aldehyde oxidises to form butanoic acid. There will be no reaction with butanone; it will remain blue.

$CH_3CH_2CH_2COH$ + Benedict's $\longrightarrow CH_3CH_2CH_2COOH$

Experiment 3: Oxidation of carbonyl compounds

Results

	Observations		
	Test 1	Test 2	Test 3
Propanal	Orange to green colour change	Silver mirror forms	Blue solution to brick-red precipitate
Propanone	No change	No change	No change
Glucose	Orange to green colour change	Silver mirror forms	Blue solution to brick-red precipitate

Questions

1 Because they both have an aldehyde functional group.

2 CH_3CH_2COH + Tollens' $\longrightarrow CH_3CH_2COOH$

3 CH_3CH_2COH + Benedict's $\longrightarrow CH_3CH_2COOH$

Checkpoint 2: Names, properties and reactions of haloalkanes, amines, aldehydes and ketones

1

Name	Structure
2-methylbutan-1-amine	$CH_3CH_2CH(CH_3)CH_2NH_2$
1-bromo-2-methylbutane	$CH_3CH_2CH(CH_3)CH_2Br$
butanal	$CH_3CH_2CH_2COH$
hexan-2-one	$CH_3C(O)(CH_2)_3CH_3$

2 Aminoethane would turn damp red litmus to blue because it is a weak base and undergoes the following reaction with water to form an alkaline solution:

$CH_3CH_2NH_2 + H_2O \rightleftharpoons CH_3CH_2NH_3^+ + OH^-$

Ethanal would, with gentle heating, turn the blue Benedict's solution to a brick-red precipitate as it is oxidised to ethanoic acid in the following reaction:

$CH_3COH + Benedict's\ solution \longrightarrow CH_3COOH$

Propanone would have no reaction with either the litmus or the Benedict's and so it would be the one left over.

3

| Name: chloroethane | $\xrightarrow{KOH_{(alc)}\ reflux}$ | Name: ethane |
| Structure: CH_3CH_2Cl | | Structure: $CH_2=CH_2$ |

Reagent and conditions: $KOH_{(alc)}$ reflux

Reagent and conditions: warm NH_3

Name: ethanamine
Structure: $CH_3CH_2NH_2$

Name: ethanol
Structure: CH_3CH_2OH

Reagent and conditions: Benedict's solution and heat

Name: ethanal
Structure: CH_3COH

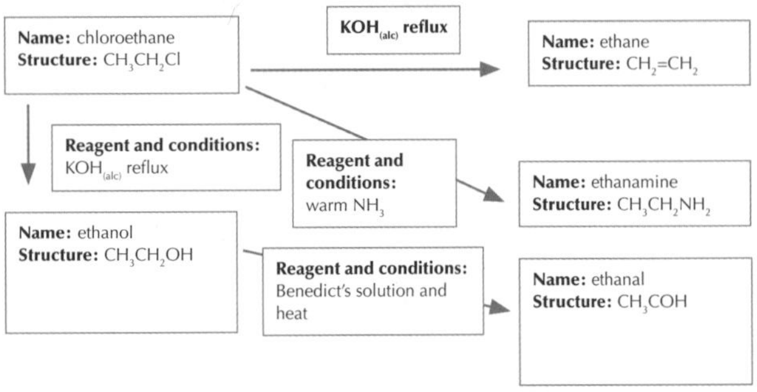

2.7: Carboxylic acids and acid chlorides

1

Name	Structure
2-methylbutanoic acid	$CH_3CH_2CH(CH_3)COOH$
3-chloro-2,3-dimethylpentanoic acid	$CH_3CH_2C(Cl)(CH_3)CH(CH_3)COOH$
3-methylbutanoyl chloride	$CH_3CH(CH_3)CH_2COCl$
butanoyl chloride	$CH_3CH_2CH_2COCl$

2 a $CH_3CH_2OH + H^+/K_2Cr_2O_7 \xrightarrow{reflux} CH_3COOH$

b $CH_3CH(CH_3)CH_2COOH + SOCl_2 \longrightarrow$ $CH_3CH(CH_3)CH_2COCl$

c $CH_3CH_2CH(CH_3)COOH + NaOH \longrightarrow$ $Na^+CH_3CH_2CH(CH_3)COO^- + H_2O$

d $2CH_3CH_2COOH + CaCO_3 \longrightarrow Ca(CH_3CH_2COO)_2 + CO_2 + H_2O$

e $CH_3COOH + CH_3CH_2CH_2OH \xrightarrow{c.\ H_2SO_4} CH_3COOCH_2CH_2CH_3$

f $CH_3CH_2COCl + CH_3OH \longrightarrow CH_3CH_2COOCH_3$

g $CH_3CH_2COOH + NH_{3\ (aq)} \xrightarrow{heat} CH_3CH_2CONH_2$

h $CH_3COCl + H_2O \longrightarrow CH_3COOH + HCl$

i $CH_3CH(CH_3)COOH + H_2O \rightleftharpoons CH_3CH(CH_3)COO^- + H_3O^+$

3 Add small amounts of water to samples of each, the one that has white vapours coming off place damp blue litmus in these vapours and they should turn red this will be ethanoyl chloride. The ethanoic acid should mix with water and will also turn the damp blue litmus red when added.

$CH_3COCl + H_2O \longrightarrow CH_3COOH + HCl$

$CH_3COOH + H_2O \rightleftharpoons CH_3COO^- + H_3O^+$

2.8: Amides and esters

1

Name	Structure
Methyl ethanoate	CH_3COOCH_3
2,2-dimethylpropanamide	$(CH_3)_3CCONH_2$
propanamide	$CH_3CH_2CONH_2$
propyl butanoate	$CH_3CH_2CH_2COOCH_2CH_2CH_3$

2 a $CH_3CH_2COOH + CH_3OH \xrightarrow{heat\ c.H_2SO_4} CH_3CH_2COOCH_3$

b $CH_3COCl + CH_3CH_2CH_2OH \longrightarrow CH_3COOCH_2CH_2CH_3$

c $CH_3CH_2COOH + NH_{3\ (aq)} \xrightarrow{heat} CH_3CH_2CONH_2$

d $CH_3CH(CH_3)COOCH_3 + dil.\ H^+ \longrightarrow CH_3CH(CH_3)COOH + CH_3OH$

e $CH_3CH(CH_3)COOCH_3 + NH_{3\ (alc)} \longrightarrow CH_3CH(CH_3)CONH_2 + CH_3OH$

f $CH_3CH(CH_3)COOCH_3 + OH^- \longrightarrow CH_3CH(CH_3)COO^- + CH_3OH$

g $CH_3CH(CH_3)CONH_2 + H_3O^+ \longrightarrow CH_3CH(CH_3)COOH + NH_4^+$

h $CH_3CH(CH_3)CONH_2 + OH^- \longrightarrow CH_3CH(CH_3)COO^- + NH_3$

Experiment 4: Preparation of esters
Questions

1 As a catalyst for the reaction and to remove water from the alcohol and carboxylic acid.

2 To neutralise any leftover sulfuric acid.

3 a $CH_3CH_2OH + CH_3COOH \longrightarrow CH_3COOCH_2CH_3 + H_2O$
Name of ester: ethyl ethanoate

b $CH_3CH_2CH_2CH_2CH_2OH + CH_3COOH \longrightarrow$ $CH_3COO(CH_2)_4CH_3 + H_2O$
Name of ester: pentyl ethanoate

c $CH_3CH_2OH + HCOOH \longrightarrow HCOOCH_2CH_3 + H_2O$
Name of ester: ethyl methanoate

d $CH_3(CH_2)_6CH_2OH + CH_3COOH \longrightarrow CH_3COO(CH_2)_7CH_3 + H_2O$
Name of ester: octyl ethanoate

e $CH_3OH + C_7H_6O_3 \longrightarrow C_7H_6O_3CH_3 + H_2O$
Name of ester: methyl salicylate

f $CH_3CH(CH_3)CH_2CH_2OH + CH_3COOH \longrightarrow$ $CH_3COOCH_2CH_2CH(CH_3)_2 + H_2O$
Name of ester: 3-methylbutyl ethanoate

2.9: Polymers and proteins

1 a $-[O - CH_2 - OOC - CH_2CH_2 - COO - CH_2 - OOC - CH_2CH_2 - COO]-$

b $-[OC - (CH_2)_{16} - CONH - CH_2 - HNOC - (CH_2)_{16} - CONH - CH_2 - NH] -$

c $-[OC - (CH_2)_{12} - CH_2OOC - CH_2 - OOC - (CH_2)_{12} - CH_2OOC - CH_2 - COO] -$

2 a $HOOCCH_2NH_2 + HOOCCH_2NH_2$

b

$HO - \overset{\overset{O}{\|}}{C} - \bigcirc - \overset{\overset{O}{\|}}{C} - O + HOCH_2CH_2OH$

c $n\ HOCH(COOH)CH_2COOH$

3 a

$H_3N^+ - \overset{\overset{H}{|}}{\underset{\underset{N - C}{|}}{C}} ...$

b

c

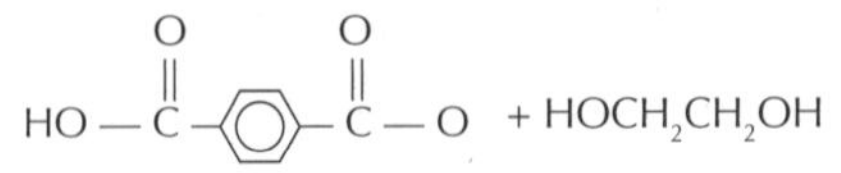

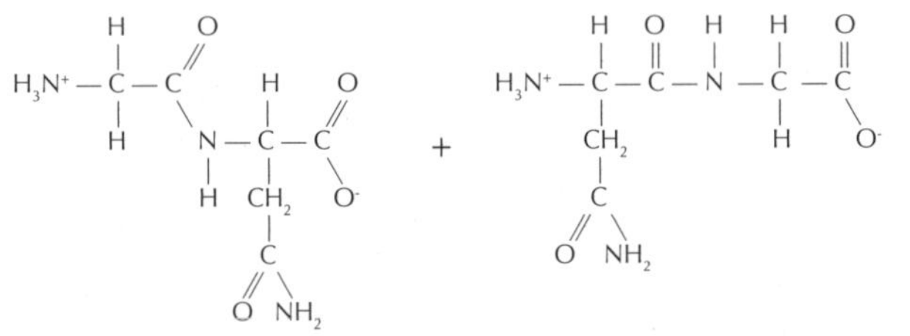

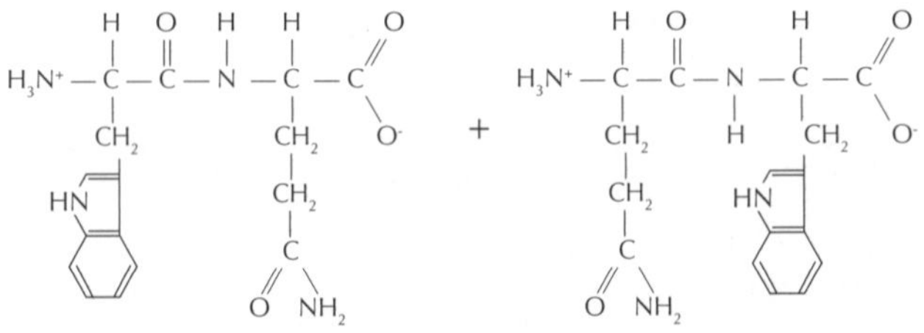

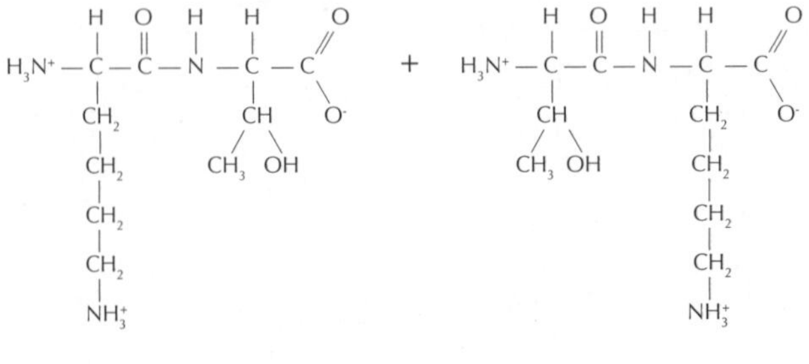

c (continued)

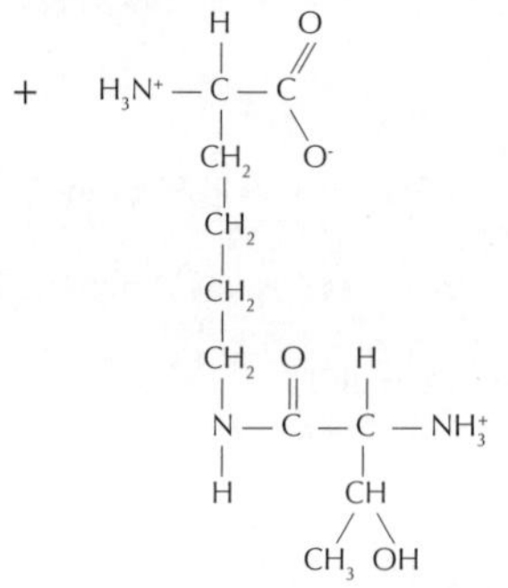

4 A condensation reaction is when a small molecule is lost from a reaction in order to form the product. As in the above cases, the small molecule of water or HCl is lost making all these condensation reactions.

Checkpoint 3: Names, properties and reactions of carboxylic acids, acid chlorides, amides and esters

1

Name	Structure
2-methylpropanoic acid	$CH_3CH(CH_3)COOH$
3-methylbutanamide	$CH_3CH(CH_3)CH_2CONH_2$
3-methylbutanoyl chloride	$CH_3CH(CH_3)CH_2COCl$
methyl propanoate	$CH_3CH_2COOCH_3$

2 The propyl methanoate will have no effect on the litmus and will only partially dissolve in water. The propanoyl chloride will release white fumes of hydrochloric acid, which will turn the damp blue litmus paper red. $CH_3CH_2COCl + H_2O \longrightarrow CH_3CH_2COOH + HCl$
The propanoic acid will turn the litmus paper red.
$CH_3CH_2COOH + H_2O \rightleftharpoons CH_3CH_2COO^- + H_3O^+$

3 Propyl methanoate contains permanent dipole-dipole bonds between its molecules due to the two oxygen atoms it contains, which require less energy to break than the hydrogen bonds between the butanamide molecules due to the H attached to the N and an O of a corresponding molecule, giving the butanamide a higher boiling point.

4

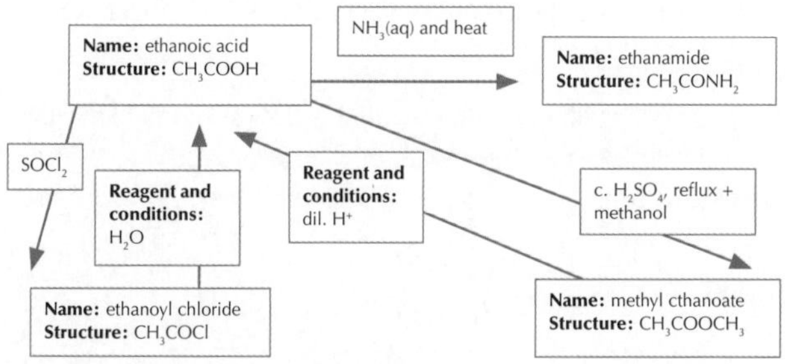

Exam-type questions

Question one

a Add acidified potassium dichromate to samples of each and reflux. The pentan-1-ol will be oxidised to pentanoic acid, which will turn damp blue litmus red and there will be a colour change of orange to green. The 2-methylbutan-2-ol will remain orange and will not be oxidised.
$CH_3CH_2CH_2CH_2CH_2OH + H^+/K_2Cr_2O_7 \longrightarrow CH_3CH_2CH_2CH_2COOH$

b Add cold acidified potassium dichromate to samples of each; the ethanol will be oxidised to ethanal and the colour will change from orange to green. The ethanoic acid will not be oxidised further and remain orange.
$CH_3CH_2OH + H^+/K_2Cr_2O_7 \longrightarrow CH_3COH$

c Add damp red litmus to samples of each; the aminopropane will turn the litmus blue as it is a weak base. The propan-1-ol will not alter the colour of litmus.
$CH_3CH_2CH_2NH_2 + H_2O \rightleftharpoons CH_3CH_2CH_2NH_3^+ + OH^-$

d Add Benedict's solution to samples of each and heat the butanal; will turn the Benedict's from a blue solution to a brick-red precipitate as it is oxidised to butanoic acid. The butanone will not be oxidised any further.
$CH_3CH_2CH_2COH$ + Benedict's/heat $\longrightarrow CH_3CH_2CH_2COOH$

e Add water to samples of each; the 1-chloropropane will barely mix with the water and the propanoyl chloride will dissolve easily to produce white vapours of hydrochloric acid.
$CH_3CH_2COCl + H_2O \longrightarrow CH_3CH_2COOH + HCl$

Question two

a

Name	Structure
propanoyl chloride	CH_3CH_2COCl
2-methylpropanal	$CH_3CH(CH_3)COH$
3-methylbutan-2-one	$CH_3C(O)CH(CH_3)_2$
2,3-dimethylbutan-1-amide	$CH_3CH(CH_3)CH(CH_3)CONH_2$
propyl propanoate	$CH_3CH_2COOCH_2CH_2CH_3$

b Propyl propanoate contains permanent dipole-dipole bonds between its molecules, which require less energy to break than the hydrogen bonds between the hexanoic acid molecules.

c 2-chloropropanal contains an asymmetric carbon, or a carbon with four different groups attached, whereas 3-chloropropanal does not contain an asymmetric carbon. Since optical isomers require an asymmetric carbon, only 2-chloropropanal can exist as optical isomers.

Question three

a

Name	Structure
2-methylpentanoic acid	$CH_3CH_2CH_2CH(CH_3)COOH$
2-methylpentanal	$CH_3CH_2CH_2CH(CH_3)COH$
2-bromopentan-3-one	$CH_3CH(Br)C(O)CH_2CH_3$
2,2-dimethylbutan-1-amine	$CH_3CH_2C(CH_3)_2CH_2NH_2$
3-methylpentanoyl chloride	$CH_3CH_2CH(CH_3)CH_2COCl$

b A physical test would be to add damp blue litmus to both and the one where it goes red is 2-methylpentanoic acid since it is a weak acid.
$CH_3CH_2CH_2CH(CH_3)COOH + H_2O \rightleftharpoons$
$CH_3CH_2CH_2CH(CH_3)COO^- + H_3O^+$
A chemical test would be to add magnesium metal. (Any metal or metal carbonate will work due to the gas being produced; you could not use a base as there will be no obvious observation to be made.) The magnesium would form bubbles of hydrogen gas in the 2-methylpentanoic acid.
$2CH_3CH_2CH_2CH(CH_3)COOH + Mg \longrightarrow$
$Mg(CH_3CH_2CH_2CH(CH_3)COO)_2 + H_2$

Question four

a **i** $CH_3CH_2CONH_2 + HCl \longrightarrow CH_3CH_2COOH + NH_4^+Cl^-$
propan-1-amide + hydrochloric acid $\longrightarrow$ propanoic acid + ammonium chloride

ii $CH_3CH_2CONH_2 + NaOH \longrightarrow CH_3CH_2COO^- + NH_3 + Na^+$
propan-1-amide + sodium hydroxide $\longrightarrow$ propanoate ion + ammonia + sodium ions

iii In a hydrolysis reaction a small molecule is added on in order to break apart the original organic structure. In both the above cases this has occurred: in the first reaction HCl was added splitting the propan-1-amide between the O and the N; in the second reaction NaOH split the propan-1-amide between the O and the N.

b **i** methyl ethanoate
ii CH_3COOCH_3 or

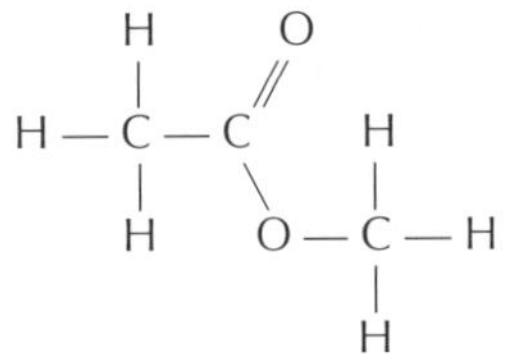

iii This reaction has two names: esterification, as it forms the ester methyl ethanoate, and also a condensation reaction because a small molecule (H_2O) is lost when the ester is formed.

Question five

a

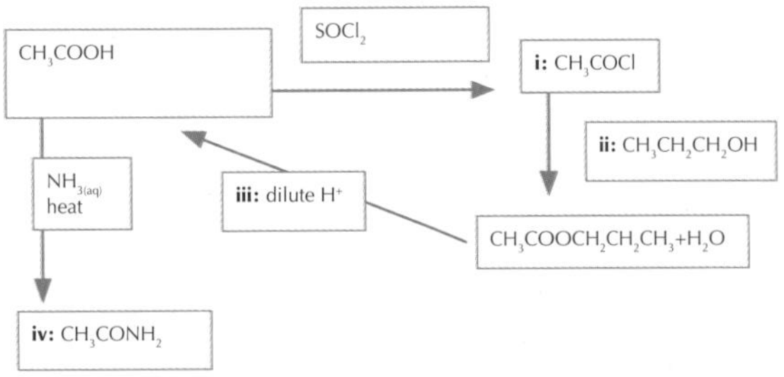

b 2-methylpropan-2-ol is a tertiary alcohol and so will not react with potassium dichromate as it cannot be oxidised. 2-methylpropan-1-ol will be oxidised with cold acidified potassium dichromate to form 2-methylpropanal.

$CH_3CH(CH_3)CH_2OH + H^+/K_2Cr_2O_7 \xrightarrow{\text{heat}} CH_3CH(CH_3)COH$
which can further be oxidised to form 2-methylpropanoic acid by adding heat.
$CH_3CH(CH_3)COH \longrightarrow CH_3CH(CH_3)COOH$
Butan-1-ol will be oxidised to form butanal using cold acidified potassium dichromate and then further oxidised to butanoic acid using heat.
$CH_3CH_2CH_2CH_2OH + H^+/K_2Cr_2O_7 \xrightarrow{\text{heat}} CH_3CH_2CH_2CH_2COH$
$\longrightarrow CH_3CH_2CH_2CH_2COOH$

Question six

Compound A is butanal, $CH_3CH_2CH_2COH$; because it reacted with Benedict's solution it had to be an aldehyde with four carbons.
$CH_3CH_2CH_2COH \longrightarrow CH_3CH_2CH_2COOH$
Compound B is butanoic acid, $CH_3CH_2CH_2COOH$, as butanal was oxidised by Benedict's to produce it.
Compound C is butan-1-ol, $CH_3CH_2CH_2CH_2OH$, as $NaBH_4$ reacted with A in the reduction equation shown below.
$CH_3CH_2CH_2COH + NaBH_4 \longrightarrow CH_3CH_2CH_2CH_2OH$
Compound D is 1-chlorobutane, $CH_3CH_2CH_2CH_2Cl$, as butan-1-ol was reacted with $SOCl_2$.
$CH_3CH_2CH_2CH_2OH + SOCl_2 \longrightarrow CH_3CH_2CH_2CH_2Cl$
Compound E is but-1-ene, $CH_2{=}CHCH_2CH_3$, as compound D underwent an elimination reaction with alcoholic KOH.
$CH_3CH_2CH_2CH_2Cl + \text{alc. } KOH \longrightarrow CH_2{=}CHCH_2CH_3$
Compound F is 2-chlorobutane, $CH_3CH(Cl)CH_2CH_3$, as it is the major product from the reaction between compound E and HCl. It is optically active because it contains a carbon with four different groups surrounding it.
$CH_2{=}CHCH_2CH_3 + HCl\ (CCl_4) \longrightarrow CH_3CH(Cl)CH_2CH_3$
Compound G is 2-aminobutane, $CH_3CH(NH_2)CH_2CH_3$, as when you warm compound F with ammonia, an amine is produced. It is optically active because it contains a carbon with four different groups surrounding it.
$CH_3CH(Cl)CH_2CH_3 + NH_3 \longrightarrow CH_3CH(NH_2)CH_2CH_3$

Question seven

Compound A is pentanoic acid, $CH_3CH_2CH_2CH_2COOH$, because it reacted with $SOCl_2$ to produce an acid chloride.
$CH_3CH_2CH_2CH_2COOH + SOCl_2 \longrightarrow CH_3CH_2CH_2CH_2COCl$
Compound B then is the acid chloride shown above, pentanoyl chloride.
Compound C is pentan-1-amide, $CH_3CH_2CH_2CH_2CONH_2$, as an acid chloride becomes an amide when alcoholic ammonia is added.
$CH_3CH_2CH_2CH_2COCl + NH_{3\,(alc)} \longrightarrow CH_3CH_2CH_2CH_2CONH_2$
Compound D is propan-1-ol, $CH_3CH_2CH_2OH$, because it reacts with A to produce an ester.
Compound E is propyl pentanoate, $CH_3CH_2CH_2CH_2COOCH_2CH_2CH_3$.
$CH_3CH_2CH_2CH_2COOH + CH_3CH_2CH_2OH + \text{c. } H_2SO_4 \longrightarrow CH_3CH_2CH_2CH_2COOCH_2CH_2CH_3$
Compound F is propene, $CH_2{=}CHCH_3$, as water is eliminated from compound D.
$CH_3CH_2CH_2OH + \text{c.}H_2SO_4 \longrightarrow CH_2{=}CHCH_3$
Compound G and H are the addition products of F reacting with HCl. G has to be 1-chloropropane as it is not optically active and H therefore must be 2-chloropropane because it contains a carbon with four different groups attached.
$CH_2{=}CHCH_3 + HCl \longrightarrow CH_3CH_2CH_2Cl + CH_3CH(Cl)CH_3$

Question eight

a **i ii** Any one of the following three dipeptides could be drawn: $H_2NCH_2COHNCH(CH_2CH_2CH_2CH_2NH_2)COOH$ or H2NCH($CH_2CH_2CH_2CH_2NH_2$)COHNCH2COOH or $H_2NCH_2COHNCH_2CH_2CH_2CH_2CH(NH_2)COOH$
iii It is a condensation reaction because a small molecule (H_2O) is lost when the dipeptide is formed.

b

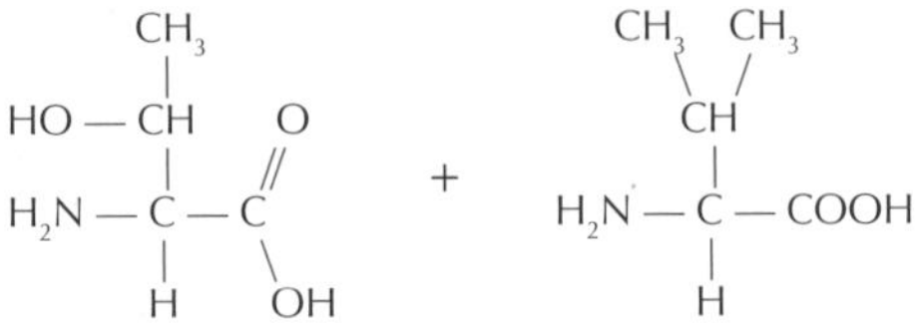

Question nine

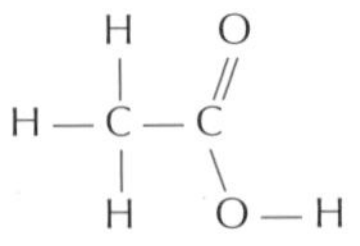

ethanoic acid

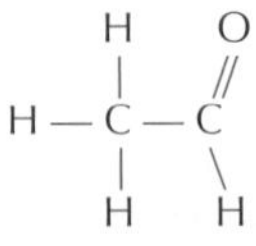

ethanal

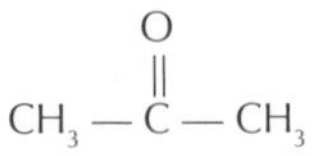

propanone

The ethanoic acid will turn damp blue litmus red as ethanoic acid is a weak acid upon the addition of water.
$CH_3COOH + H_2O \rightleftharpoons CH_3COO^- + H_3O^+$
The ethanal will react with Benedict's solution to form ethanoic acid, and there will be a colour change from blue to a brick-red precipitate when this occurs.
$CH_3COH + \text{Benedict's/heat} \longrightarrow CH_3COOH$
The propanone will mix with water but will show no change with either Benedict's or litmus.

Question ten

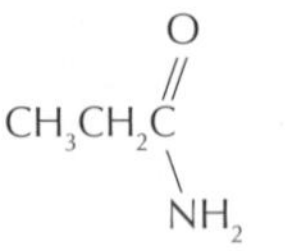

propanamide

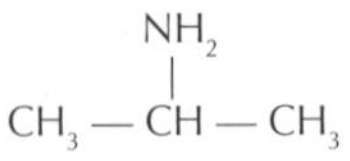

propan-2-amine

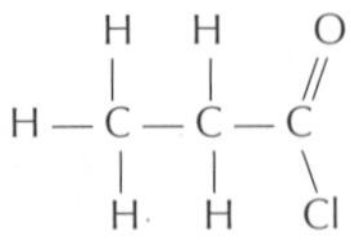

propanoyl chloride

The propanamide will show no reaction with litmus but will dissolve in water.
The propan-2-amine will also mix with the water but it will turn damp red litmus blue because it is a weak base.
$CH_3CH(NH)CH_3 + H_2O \rightleftharpoons CH_3CH(NH_3^+)CH_3 + OH^-$
The propanoyl chloride will form white fumes of hydrochloric acid when mixed with water, which will turn damp blue litmus red.
$CH_3CH_2COCl + H_2O \longrightarrow HCl + CH_3CH_2COOH$

Chapter Three

Pre-test

1 **a** $K_c = \dfrac{[PCl_5]}{[PCl_3][Cl_2]}$

ISBN: 9780170352611

b $K_c = \frac{[O_2][NO_2]^4}{[N_2O_5]^2}$

c $K_c = \frac{[HSO_3^-][OH^-]}{[SO_3^{2-}]}$

d $K_c = \frac{[CO_2][H_2]}{[CO][H_2O]}$

2 **a** Increasing the temperature will force the equilibrium system to shift in the endothermic direction in order to lower the increased temperature; in this case this will be towards the reactants as the forward reaction is exothermic. K_c will lower in value as there will be more reactants to products in the new equilibrium that is established.

b Increasing the total pressure will force the equilibrium to shift to the side with the fewest number of moles in order to decrease the pressure of the system; in this case this is towards the reactants as there are 9 moles on this side compared with the 10 on the products side. K_c will not be affected by this change.

c Adding a catalyst will increase both the forward and reverse reactions equally and so will have no effect on the position of the equilibrium or the value of K_c.

d Decreasing the ammonia, NH_3, concentration will force the equilibrium to shift towards the reactants in order to replace the ammonia that was lost. K_c will remain unaltered as changes in concentration have no effect on the value of K_c.

3 **a** $HCl + H_2O \longrightarrow Cl^- + H_3O^+$

b $H_2SO_4 + 2H_2O \longrightarrow SO_4^{2-} + 2H_3O^+$ or $H_2SO_4 + H_2O \longrightarrow HSO_4^- + H_3O^+$

c $CH_3CH_2COOH + H_2O \rightleftharpoons CH_3CH_2COO^- + H_3O^+$

d $Mg(OH)_{2\,(aq)} \longrightarrow Mg^{2+} + 2OH^-$

e $NH_3 + H_2O \rightleftharpoons NH_4^+ + OH^-$

4 **a** 1 $molL^{-1}$ hydrochloric acid means there is the same concentration of hydronium ions since it fully dissociates.
$pH = -log[H_3O^+] = -log[1] = 0$

b 0.0122 $molL^{-1}$ hydrochloric acid means there is the same concentration of hydronium ions since it fully dissociates.
$pH = -log[H_3O^+] = -log[0.0122] = 1.91$ (3 s.f.)

5 **a** $[H_3O^+] = K_w/[OH^-]$. Since $[OH^-] = [NaOH] = 0.100\ molL^{-1}$
$[H_3O^+] = 10^{-14}/0.100 = 1.00 \times 10^{-13}\ molL^{-1}$
$pH = -log[H_3O^+] = 13.0$ (3 s.f.)

b $[H_3O^+] = K_w/[OH^-]$. Since $[OH^-] = [NaOH] = 1.20\ molL^{-1}$
$[H_3O^+] = 10^{-14}/1.20 = 8.333... \times 10^{-15}\ molL^{-1}$
$pH = -log[H_3O^+] = 14.1$ (3 s.f.)

6 Propanoic acid is a weak acid and so will only partially dissociate in water to produce hydronium ions and its conjugate base propanoate ions.
$CH_3CH_2COOH_{(aq)} + H_2O_{(l)} \rightleftharpoons CH_3CH_2COO^-_{(aq)} + H_3O^+_{(aq)}$
Therefore there will be a lower concentration of ions in solution able to conduct a current and since ions are required to conduct a current it will be a weaker conductor than sulfuric acid. Sulfuric acid is a strong acid and it also has two protons that it can protonate therefore it will completely dissociate in water leaving more ions in solution and so it will be a better conductor of electricity.
$H_2SO_{4\,(aq)} + 2H_2O_{(l)} \longrightarrow SO_4^{2-}{}_{(aq)} + 2H_3O^+_{(aq)}$

Experiment 1: Solubility

Method and results

1 A yellow precipitate of lead iodide was formed.

2 The yellow precipitate dissolved and yellow crystals of lead iodide were left behind.

Conclusion

When you heat an insoluble substance you force it dissolve again, which must mean that an insoluble substance is in equilibrium with its ions and therefore it is a reversible reaction.
$PbI_{2\,(s)} \rightleftharpoons Pb^{2+}_{(aq)} + 2I^-_{(aq)}$

3.1: Solubility equilibria

Compound	Formula	Equation	K_s expression
Silver iodide	AgI	$AgI_{(s)} \rightleftharpoons Ag^+_{(aq)} + I^-_{(aq)}$	$K_s = [Ag^+][I^-]$
Zinc sulfide	ZnS	$ZnS_{(s)} \rightleftharpoons Zn^{2+}_{(aq)} + S^{2-}_{(aq)}$	$K_s = [Zn^{2+}][S^{2-}]$
Copper (II) chloride	$CuCl_2$	$CuCl_{2\,(s)} \rightleftharpoons Cu^{2+}_{(aq)} + 2Cl^-_{(aq)}$	$K_s = [Cu^{2+}][Cl^-]^2$
Calcium fluoride	CaF_2	$CaF_{(s)} \rightleftharpoons Ca^{2+}_{(aq)} + 2F^-_{(aq)}$	$K_s = [Ca^{2+}][F^-]^2$
Lead iodide	PbI_2	$PbI_{2\,(s)} \rightleftharpoons Pb^{2+}_{(aq)} + 2I^-_{(aq)}$	$K_s = [Pb^{2+}][I^-]^2$
Iron (II) sulfate	$FeSO_4$	$FeSO_{4\,(s)} \rightleftharpoons Fe^{2+}_{(aq)} + SO_4^{2-}{}_{(aq)}$	$K_s = [Fe^{2+}][SO_4^{2-}]$
Silver carbonate	Ag_2CO_3	$Ag_2CO_{3\,(s)} \rightleftharpoons 2Ag^+_{(aq)} + CO_3^{2-}{}_{(aq)}$	$K_s = [Ag^+]^2[CO_3^{2-}]$
Lead chloride	$PbCl_2$	$PbCl_{2\,(s)} \rightleftharpoons Pb^{2+}_{(aq)} + 2Cl^-_{(aq)}$	$K_s = [Pb^{2+}]Cl^-]^2$

3.2: Solubility product calculations

2 $AgCl_{(s)} \rightleftharpoons Ag^+_{(aq)} + Cl^-_{(aq)}$
$K_s = [Ag^+][Cl^-]$
$K_s = s^2 = (1.60 \times 10^{-3})^2 = 2.56 \times 10^{-6}$ (3 s.f.)

3 $PbI_{2\,(s)} \rightleftharpoons Pb^{2+}_{(aq)} + 2I^-_{(aq)}$
$K_s = [Pb^{2+}][I^-]^2$
$K_s = 4s^3 = 4 \times (1.51 \times 10^{-4})^3 = 1.38 \times 10^{-11}$ (3 s.f.)

4 $Ag_2SO_{4\,(s)} \rightleftharpoons 2Ag^+_{(aq)} + SO_4^{2-}{}_{(aq)}$
$K_s = [Ag^+]^2[SO_4^{2-}]$
$K_s = 4s^3 = 4 \times (1.31 \times 10^{-5})^3 = 8.99 \times 10^{-15}$ (3 s.f.)

5 $CuCO_{3\,(s)} \rightleftharpoons Cu^{2+}_{(aq)} + CO_3^{2-}{}_{(aq)}$
$K_s = [Cu^{2+}][CO_3{}^{2-}]$
$K_s = s^2 = (1.2 \times 10^{-5})^2 = 1.44 \times 10^{-10}$ (3 s.f.)

6 $BaSO_{4\,(s)} \rightleftharpoons Ba^{2+}_{(aq)} + SO_4^{2-}{}_{(aq)}$
$K_s = [Ba^{2+}][SO_4^{2-}]$
$K_s = s^2 \quad s = \sqrt{K_s} = \sqrt{1.1 \times 10^{-10}} = 1.05 \times 10^{-5}\ molL^{-1}$ (3 s.f.)

7 $AgI_{(s)} \rightleftharpoons Ag^+_{(aq)} + I^-_{(aq)}$
$K_s = [Ag^+][I^-]$
$K_s = s^2 \quad s = \sqrt{K_s} = \sqrt{8.3 \times 10^{-17}} = 9.11 \times 10^{-9}\ molL^{-1}$ (3 s.f.)

8 $Cu(OH)_{2\,(s)} \rightleftharpoons Cu^{2+}_{(aq)} + 2OH^-_{(aq)}$
$K_s = [Cu^{2+}][OH^-]^2$
$K_s = 4s^3 \quad s = \sqrt[3]{K_s/4} = \sqrt[3]{4.8 \times 10^{-20}/4} = 2.29 \times 10^{-7}\ molL^{-1}$ (3 s.f.)

9 $CaF_{2\,(s)} \rightleftharpoons Ca^{2+}_{(aq)} + 2F^-_{(aq)}$
$K_s = [Ca^{2+}][F^-]^2$
$K_s = 4s^3 \quad s = \sqrt[3]{K_s/4} = \sqrt[3]{3.45 \times 10^{-11}/4} = 2.05 \times 10^{-4}\ molL^{-1}$ (3 s.f.)

10 $MgF_{2\,(S)} \rightleftharpoons Mg^{2+}_{(aq)} + 2F^-_{(aq)}$
$K_s = [Mg^{2+}][F^-]^2$
$K_s = 4s^3 \quad s = \sqrt[3]{K_s/4} = \sqrt[3]{5.16 \times 10^{-11}/4} = 2.35 \times 10^{-4}\ molL^{-1}$ (3 s.f.)

3.3: The common ion effect

1 $BaCO_{3\,(s)} \rightleftharpoons Ba^{2+}_{(aq)} + CO_3^{2-}{}_{(aq)}$
$K_s = [Ba^{2+}][CO_3^{2-}] = s \times (s + 0.300)$
The assumption is s is much less than 0.300 $molL^{-1}$
$K_s = s \times 0.300$
$s = K_s/0.300 = 5 \times 10^{-9}/0.300 = 1.67 \times 10^{-8}\ molL^{-1}$ (3 s.f.)

2 $PbSO_{4\,(s)} \rightleftharpoons Pb^{2+}_{(aq)} + SO_4^{2-}{}_{(aq)}$
$K_s = [Pb^{2+}][SO_4^{2-}] = s \times (s + 0.0501)$
The assumption is s is much less than 0.0501 $molL^{-1}$
$s = K_s/0.0501 = 2 \times 10^{-8}/0.0501 = 3.99 \times 10^{-7}\ molL^{-1}$

3 $CaF_{2\,(s)} \rightleftharpoons Ca^{2+}_{(aq)} + 2F^-_{(aq)}$
$K_s = [Ca^{2+}][F^-]^2 = (s + 0.100) \times (2s)^2$
The assumption is s is much less than 0.100 $molL^{-1}$
$s = \sqrt{K_s/(4 \times 0.100)} = \sqrt{4 \times 10^{-11}/(4 \times 0.100)} = 1 \times 10^{-5}\ molL^{-1}$

4 $PbI_{2\,(s)} \rightleftharpoons Pb^{2+}_{(aq)} + 2I^-_{(aq)}$
$K_s = [Pb^{2+}][I^-]^2 = s \times (s + 0.101)^2$
The assumption is s is much less than 0.0101 $molL^{-1}$
$s = K_s/0.0101^2 = 1 \times 10^{-9}/0.0101^2 = 9.80 \times 10^{-6}\ molL^{-1}$ (3 s.f.)

5 $AgI_{(s)} \rightleftharpoons Ag^+_{(aq)} + I^-_{(aq)}$
$K_s = [Ag^+][I^-] = (s + 1.10 \times 10^{-3}) \times s$
Assumption that s is much smaller than $1.10 \times 10^{-3}\ molL^{-1}$
$s = K_s/1.10 \times 10^{-3} = 8 \times 10^{-17}/1.10 \times 10^{-3} = 7.27 \times 10^{-14}\ molL^{-1}$ (3 s.f.)

Checkpoint 1: Solubility and the common ion effect

1 **a** $K_s(AlPO_4) = 9.84 \times 10^{-21}$
$AlPO_{4\,(s)} \rightleftharpoons Al^{3+}_{(aq)} + PO_4^{3-}{}_{(aq)}$
$K_s = [Al^{3+}][PO_3^{4-}]$
$K_s = s^2$
$s = \sqrt{K_s} = \sqrt{9.84 \times 10^{-21}} = 9.92 \times 10^{-11}\ molL^{-1}$

b $K_s(BaF_2) = 1.84 \times 10^{-7}$
$BaF_{2\,(s)} \rightleftharpoons Ba^{2+}{}_{(aq)} + 2F^-{}_{(aq)}$
$K_s = [Ba^{2+}][F^-]^2$
$K_s = 4s^3$
$s = \sqrt[3]{K_s/4} = \sqrt[3]{1.84 \times 10^{-7}/4} = 0.00358\ molL^{-1}$

c $K_s(PbBr_2) = 6.60 \times 10^{-6}$
$PbBr_{2\,(s)} \rightleftharpoons Pb^{2+}{}_{(aq)} + 2Br^-{}_{(aq)}$
$K_s = [Pb^{2+}][Br^-]^2$
$K_s = 4s^3$
$s = \sqrt[3]{K_s/4} = \sqrt[3]{6.60 \times 10^{-6}/4} = 0.0118\ molL^{-1}$

d $K_s(KIO_4) = 3.71 \times 10^{-4}$
$KIO_{4\,(s)} \rightleftharpoons K^+{}_{(aq)} + IO_4^-{}_{(aq)}$
$K_s = [K^+][IO_4^-]$
$K_s = s^2$
$s = \sqrt{K_s} = \sqrt{3.71 \times 10^{-4}} = 0.0193\ molL^{-1}$

e $K_s(Sn(OH)_2) = 5.45 \times 10^{-27}$
$Sn(OH)_{2\,(s)} \rightleftharpoons Sn^{2+}{}_{(aq)} + 2OH^-{}_{(aq)}$
$K_s = [Sn^{2+}][OH^-]^2$
$K_s = 4s^3$
$s = \sqrt[3]{K_s/4} = \sqrt[3]{5.45 \times 10^{-27}/4} = 1.11 \times 10^{-9}\ molL^{-1}$

2 a For $ZnCO_3$ where $s = 1.21\times10^{-5}\ molL^{-1}$
$ZnCO_{3\,(s)} \rightleftharpoons Zn^{2+}{}_{(aq)} + CO_3^{2-}{}_{(aq)}$
$K_s = [Zn^{2+}][CO_3^{2-}] = s^2 = (1.21 \times 10^{-5})^2 = 1.46 \times 10^{-10}$

b For $Ni(OH)_2$ where $s = 5.16 \times 10^{-6}\ molL^{-1}$
$Ni(OH)_{2\,(s)} \rightleftharpoons Ni^{2+}{}_{(aq)} + 2OH^-{}_{(aq)}$
$K_s = [Ni^{2+}][OH^-]^2 = 4s^3 = 4(5.16 \times 10^{-6})^3 = 5.50 \times 10^{-16}$

c For HgI_2 where $s = 1.94 \times 10^{-10}\ molL^{-1}$
$HgI_{2\,(s)} \rightleftharpoons Hg^{2+}{}_{(aq)} + 2I^-{}_{(aq)}$
$K_s = [Hg^{2+}][I^-]^2 = 4s^3 = 4(1.94 \times 10^{-10})^3 = 2.92 \times 10^{-29}$

d For Ag_2CO_3 where $s = 1.28 \times 10^{-4}\ molL^{-1}$
$Ag_2CO_{3\,(s)} \rightleftharpoons 2Ag^+{}_{(aq)} + CO_3^{2-}{}_{(aq)}$
$K_s = [Ag^+]^2[CO_3^{2-}] = 4s^3 = 4(1.28 \times 10^{-4})^3 = 8.39 \times 10^{-12}$

e For $SrSO_4$ where $s = 5.87 \times 10^{-4} molL^{-1}$
$SrSO_{4\,(s)} \rightleftharpoons Sr^{2+}{}_{(aq)} + SO_4^{2-}{}_{(aq)}$
$K_s = [Sr^{2+}][SO_4^{2-}] = s^2 = (5.87 \times 10^{-4})^2 = 3.45 \times 10^{-7}$

3 a $K_s(HgS) = 2 \times 10^{-53}$
$c(Na_2S) = 0.105\ molL^{-1}$
$HgS_{(s)} \rightleftharpoons Hg^{2+}(aq) + S^{2-}{}_{(aq)}$
$K_s = [Hg^{2+}][S^{2-}] = s \times (s + 0.105)$
We assume that s is much less than 0.105 $molL^{-1}$
$s = K_s/0.105 = 2 \times 10^{-53}/0.105 = 1.90 \times 10^{-52}\ molL^{-1}$

b $K_s(MnCO_3) = 2.24 \times 10^{-11}$
$c(Na_2CO_3) = 0.166\ molL^{-1}$
$MnCO_{3\,(s)} \rightleftharpoons Mn^{2+}(aq) + CO_3^{2-}{}_{(aq)}$
$K_s = [Mn^{2+}][CO_3^{2-}] = s \times (s + 0.166)$
We assume that s is much less than 0.166 $molL^{-1}$
$s = K_s/0.166 = 2.24 \times 10^{-11}/0.166 = 1.35 \times 10^{-10}\ molL^{-1}$

c $K_s(FeF_2) = 2.36 \times 10^{-6}$
$c(Fe(NO_3)_2) = 0.201\ molL^{-1}$
$FeF_{2\,(s)} \rightleftharpoons Fe^{2+}(aq) + 2F^-{}_{(aq)}$
$K_s = [Fe^{2+}][F^-]^2 = (s + 0.201) \times (2s)^2$
We assume that s is much less than 0.201 $molL^{-1}$
$s = \sqrt{K_s/(0.201 \times 4)} = \sqrt{2.36 \times 10^{-6}/(0.201 \times 4)} = 1.71 \times 10^{-3}$ $molL^{-1}$

d $K_s(Cu(OH)_2) = 4.8 \times 10^{-20}$
$c(KOH) = 1.22\ molL^{-1}$
$Cu(OH)_{2\,(s)} \rightleftharpoons Cu^{2+}(aq) + 2OH^-{}_{(aq)}$
$K_s = [Cu^{2+}][OH^-]^2 = s \times (s + 0.122)^2$
We assume that s is much less than 0.122 $molL^{-1}$
$s = K_s/0.122^2 = 4.8 \times 10^{-20}/0.122^2 = 3.22 \times 10^{-20}\ molL^{-1}$

e $K_s(CdF_2) = 6.44 \times 10^{-3}$
$c(KF) = 0.0607\ molL^{-1}$
$CdF_{2\,(s)} \rightleftharpoons Cd^{2+}(aq) + 2F^-{}_{(aq)}$
$K_s = [Cd^{2+}][F^-]^2 = s \times (s + 0.0607)^2$
We assume that s is much less than 0.105 $molL^{-1}$
$s = K_s/0.0607^2 = 6.44 \times 10^{-3}/0.0607^2 = 1.78\ molL^{-1}$

3.4: Predicting precipitation

a $CuS_{(s)} \rightleftharpoons Cu^{2+}{}_{(aq)} + S^{2-}{}_{(aq)}$
$Q_s = [Cu^{2+}][S^{2-}] = (5.61 \times 10^{-3} \times 55(155 + 55)) \times (6.11 \times 10^{-5} \times 155(55 + 155)) = 6.63 \times 10^{-8}$
$Q_s > K_s$ so a precipitate will occur

b $FeF_{2\,(s)} \rightleftharpoons Fe^{2+}{}_{(aq)} + 2F^-{}_{(aq)}$
$Q_s = [Fe^{2+}][F^-]^2 = (0.0501 \times 65(65 + 45)) \times (0.00102 \times 45(65 + 45))^2 = 5.15 \times 10^{-9}$
$Q_s < K_s$ so no precipitate will occur

c $MgCO_{3\,(s)} \rightleftharpoons Mg^{2+}{}_{(aq)} + CO_3^{2-}{}_{(aq)}$
$Q_s = [Mg^{2+}][CO_3^{2-}] = (0.000601 \times 45(45 + 45)) \times (0.000702 \times 45(45 + 45)) = 1.05 \times 10^{-7}$
$Q_s < K_s$ so no precipitate will occur

d $AgCl_{(s)} \rightleftharpoons Ag^+{}_{(aq)} + Cl^-{}_{(aq)}$
$Q_s = [Ag^+][Cl^-] = (0.00101 \times 75(75 + 145)) \times (0.00602 \times 145(75 + 145)) = 1.37 \times 10^{-6}$
$Q_s > K_s$ so a precipitate will occur

e $Zn(OH)_{2\,(s)} \rightleftharpoons Zn^{2+}{}_{(aq)} + 2OH^-{}_{(aq)}$
$Q_s = [Zn^{2+}][OH^-]^2 = (0.100 \times 65(65 + 155)) \times (0.0202 \times 155(65 + 155))^2 = 5.98 \times 10^{-6}$
$Q_s > K_s$ so a precipitate will occur

f $CoCO_{3\,(s)} \rightleftharpoons Co^{2+}{}_{(aq)} + CO_3^{2-}{}_{(aq)}$
$Q_s = [Co^{2+}][CO_3^{2-}] = (0.000502 \times 100(100 + 57)) \times (0.000141 \times 57(155 + 57)) = 1.64 \times 10^{-8}$
$Q_s > K_s$ so a precipitate will occur

g $PbI_{2\,(s)} \rightleftharpoons Pb^{2+}{}_{(aq)} + 2I^-{}_{(aq)}$
$Q_s = [Pb^{2+}][I^-]^2 = (0.00301 \times 165(165 + 175)) \times (0.00607 \times 175(165 + 175))^2 = 1.44 \times 10^{-8}$
$Q_s > K_s$ so a precipitate will occur

Experiment 2: Complex ions

Method and results

1

$AgNO_3$ + NaCl; observations and name of precipitate formed	AgCl + NH_3; observations and formula of precipitate formed
A white precipitate of silver chloride, AgCl, forms	The white precipitate disappears leaving a colourless solution of the complex ion $[Ag(NH_3)_2]^+$

2

$Cu(NO_3)_2$ + NH_3; observations and name of precipitate formed	$Cu(OH)_2$ + excess NH_3; observations and formula of precipitate formed
A blue precipitate of copper hydroxide, $Cu(OH)_2$, forms	The precipitate disappears leaving a deep-blue solution of $[Cu(NH_3)_4]^{2+}$

3

$Zn(NO_3)_2$ + NaOH; observations and name of precipitate formed	$Zn(OH)_2$ + excess NaOH; observations and formula of precipitate formed
A white precipitate of zinc hydroxide, $Zn(OH)_2$, forms	The precipitate disappears leaving a colourless solution of $[Zn(OH)_4]^{2-}$

4

$Zn(NO_3)_2$ + NH_3; observations and name of precipitate formed	$Zn(OH)_2$ + excess NH_3; observations and formula of precipitate formed
A white precipitate of zinc hydroxide, $Zn(OH)_2$, forms	The precipitate disappears leaving a colourless solution of $[Zn(NH_3)_4]^{2+}$

5

$Pb(NO_3)_2$ + NaOH; observations and name of precipitate formed	$Pb(OH)_2$ + excess NaOH; observations and formula of precipitate formed
A white precipitate of lead hydroxide, $Pb(OH)_2$, forms	The precipitate disappears leaving a colourless solution of $[Pb(OH)_4]^{2-}$

Equations

1 $Ag^+{}_{(aq)} + Cl^-{}_{(aq)} \rightarrow AgCl_{(s)}$
$AgCl_{(s)} + 2NH_{3\,(aq)} \rightarrow [Ag(NH_3)_2]^+{}_{(aq)} + Cl^-{}_{(aq)}$

ISBN: 9780170352611

2 $Cu^{2+}_{(aq)} + 2OH^-_{(aq)} \longrightarrow Cu(OH)_{2\,(s)}$
$Cu(OH)_{2\,(s)} + 4NH_{3\,(aq)} \longrightarrow [Cu(NH_3)_4]^{2+}_{(aq)} + 2OH^-_{(aq)}$

3 $Zn^{2+}_{(aq)} + 2OH^-_{(aq)} \longrightarrow Zn(OH)_{2\,(s)}$
$Zn(OH)_{2\,(s)} + 2OH^-_{(aq)} \longrightarrow [Zn(OH)_4]^{2-}_{(aq)}$

4 $Zn^{2+}_{(aq)} + 2OH^-_{(aq)} \longrightarrow Zn(OH)_{2\,(s)}$
$Zn(OH)_{2\,(s)} + 4NH_{3\,(aq)} \longrightarrow [Zn(NH_3)_4]^{2+}_{(aq)} + 2OH^-_{(aq)}$

5 $Pb^{2+}_{(aq)} + 2OH^-_{(aq)} \longrightarrow Pb(OH)_{2\,(s)}$
$Pb(OH)_{2\,(s)} + 2OH^-_{(aq)} \longrightarrow [Pb(OH)_4]^{2-}_{(aq)}$

3.5: Complex ions and solubility

1 The solubility of AgCl will increase as the equilbirium reaction $AgCl_{(s)} \rightleftharpoons Ag^+_{(aq)} + Cl^-_{(aq)}$ will be forced towards the products as the silver ions are taken out by forming the complex ion $[Ag(NH_3)_2]^+$. The formation of this complex can be represented by the following equation:
$AgCl_{(s)} + 2NH_{3\,(aq)} \longrightarrow [Ag(NH_3)_2]^+_{(aq)} + Cl^-_{(aq)}$

2 The blue precipitate that forms is copper hydroxide which forms in ammonia solution as ammonia is a weak base and so has hydroxide ions present in solution; this is represented in the following equation:
$Cu^{2+}_{(aq)} + 2OH^-_{(aq)} \longrightarrow Cu(OH)_{2\,(s)}$
In excess ammonia this blue precipitate disappears and a deep-blue solution is the result, which can be represented by the following reaction:
$Cu(OH)_{2\,(s)} + 4NH_{3\,(aq)} \longrightarrow [Cu(NH_3)_4]^{2+}_{(aq)} + 2OH^-_{(aq)}$
This happens because the solubility of the copper hydroxide is increased upon the addition of ammonia due to the formation of the complex ion with the copper ions.

3 A precipitate will form when K_s equals Q_s, so:
$Q_s = [Zn^{2+}][OH^-]^2$
$3 \times 10^{-17} = 0.1 \times [OH^-]^2$
$[OH^-] = \sqrt{3 \times 10^{-17}/0.1} = 1.73 \times 10^{-8}$ molL^{-1}
$K_w = [H_3O^+][OH^-]$
$[H_3O^+] = K_w/[OH^-] = 1 \times 10^{-14}/1.73 \times 10^{-8} = 5.77... \times 10^{-7}$ molL^{-1}
pH = $-\log[H_3O^+] = 6.24$

If the pH is raised then the solubility will increase due to the formation of the zincate complex ion:
$Zn(OH)_{2\,(s)} + 2OH^-_{(aq)} \longrightarrow [Zn(OH)_4]^{2-}_{(aq)}$
If the pH is lowered there will be a lower concentration of hydroxide ions in solution making zinc hydroxide more soluble as any acid particles will react with the hydroxide ions removing them from the equilibrium.

4 Adding acid will force the equilibrium towards the products as the acid particles will react with the hydroxide ions and remove them from the equilibrium.
$Pb(OH)_{2\,(s)} \rightleftharpoons Pb^{2+}_{(aq)} + 2OH^-_{(aq)}$
$OH^- + H_3O^+ \longrightarrow H_2O$
Adding base will force the equilibrium to shift towards the products as well as the increased hydroxide ions will form a complex ion with the lead hydroxide:
$Pb(OH)_{2\,(s)} + 2OH^-_{(aq)} \longrightarrow [Pb(OH)_4]^{2-}_{(aq)}$

3.6: Species in solution

Solution	Species present in solution
HNO_3	$H_3O^+ = NO_3^- >>> OH^-$
HBr	$H_3O^+ = Br >>> OH^-$
HF	$HF > H_3O^+ = F^- >> OH^-$
$Mg(OH)_2$	$OH^- > Mg^{2+} >> H_3O^+$
CH_3COO^-	$CH_3COO^- > OH^- = CH_3COOH >> H_3O^+$

Checkpoint 2: Predicting precipitation and complex ions

1 a $Q_s = [Ca^{2+}][SO_4^{2-}] = (2.01 \times 10^{-4} \times 40(40 + 60)) \times (1.06 \times 10^{-5} \times 60(40 + 60)) = 5.11 \times 10^{-10}$
$Q_s < K_s$ so no precipitate will form

b $Q_s = [Hg^{2+}][I^-]^2 = (1.66 \times 10^{-3} \times 55(55 + 65)) \times (1.06 \times 10^{-3} \times 65(55 + 65))^2 = 2.51 \times 10^{-10}$
$Q_s > K_s$ so a precipitate will form

c $Q_s = [Ni^{2+}][CO_3^{2-}] = (6 \times 10^{-6} \times 115(115 + 85)) \times (1.11 \times 10^{-3} \times 85(115 + 85)) = 1.63 \times 10^{-9}$
$Q_s < K_s$ so no precipitate will form

d $Q_s = [Ag^+][CH_3COO^-] = (0.2 \times 15(15 + 35)) \times (0.1 \times 35(35 + 15)) = 0.00420$
$Q_s > K_s$ so a precipitate will form

2 a When ammonia is added to $Cu(OH)_2$ solution the solubility will increase as a complex ion will form with the copper ions and ammonia molecules making a deep-blue solution.

b When ammonia is added to $Zn(OH)_2$ solution the solubility will increase as a complex ion will form with the zinc ions and ammonia molecules removing them from the equilibrium.

c The sodium hydroxide will form a complex ion with the $Zn(OH)_2$ making the solubility of the zinc hydroxide increase as more and more of the zinc ions will be removed from the equilibrium.

3 a $H_3O^+ > SO_4^{2-} >>> OH^-$

b $K^+ = OH^- >> H_3O^+$

c $NH_3 > NH_4^+ = OH^- > > H_3O^+$

d $NH_4^+ > NH_3 = H_3O^+ >> OH^-$

3.7: Calculating the pH of acids

1 a $HCl_{(aq)} + H_2O_{(l)} \longrightarrow Cl^-_{(aq)} + H_3O^+_{(aq)}$
pH = $-\log[H_3O^+] = -\log 0.01 = 2.00$

b $HBr_{(aq)} + H_2O_{(l)} \longrightarrow Br^-_{(aq)} + H_3O^+_{(aq)}$
pH = $-\log[H_3O^+] = -\log 1.03 = -0.0128$

c $HNO_{3\,(aq)} + H_2O_{(l)} \longrightarrow NO_3^-{}_{(aq)} + H_3O^+_{(aq)}$
pH = $-\log[H_3O^+] = -\log 0.126 = 0.900$

2 a $CH_3COOH_{(aq)} + H_2O_{(l)} \rightleftharpoons CH_3COO^-_{(aq)} + H_3O^+_{(aq)}$
$K_a = \dfrac{[CH_3COO^-][H_3O^+]}{[CH_3COOH]}$
We assume that $[CH_3COO^-] = [H_3O^+]$ and the initial concentration of $[CH_3COOH]$ remains the same
$[H_3O^+] = \sqrt{K_a \times [CH_3COOH]} = \sqrt{1.8 \times 10^{-5} \times 0.01} = 4.24... \times 10^{-4}$ molL^{-1}
pH = $-\log[H_3O^+] = 3.37$

b $HF_{(aq)} + H_2O\ (l) \rightleftharpoons F^-_{(aq)} + H_3O^+_{(aq)}$
$K_a = \dfrac{[F^-][H_3O^+]}{[HF]}$
We assume that $[F^-] = [H_3O^+]$ and the initial concentration of [HF] remains the same
$[H_3O^+] = \sqrt{K_a \times [HF]} = \sqrt{6.3 \times 10^{-4} \times 0.01} = 0.00250...$ molL^{-1}
pH = $-\log[H_3O^+] = 2.60$

c $CH_3CH_2COOH_{(aq)} + H_2O_{(l)} \rightleftharpoons CH_3CH_2COO^-_{(aq)} + H_3O^+_{(aq)}$
$K_a = \dfrac{[CH_3CH_2COO^-][H_3O^+]}{[CH_3CH_2COOH]}$
We assume that $[CH_3CH_2COO^-] = [H_3O^+]$ and the initial concentration of $[CH_3CH_2COOH]$ remains the same
$[H_3O^+] = \sqrt{K_a \times [CH_3CH_2COOH]} = \sqrt{1.3 \times 10^{-5} \times 0.01} = 3.60... \times 10^{-4}$ molL^{-1}
pH = $-\log[H_3O^+] = 3.44$

3.8: pK_a and K_a

1 a 4.89
b 3.20
c 4.74

2 a 3.98×10^{-4}
b 3.16×10^{-8}
c 6.31×10^{-10}
Order of strength: $HNO_2 > HOCl > HCN$

3.9: Calculate the pH of the following bases and explain their conductivity

1 a $NaOH_{(aq)} \longrightarrow Na^+_{(aq)} + OH^-_{(aq)}$
$[H_3O^+] = K_w/[OH^-] = 1 \times 10^{-14}/0.01 = 1 \times 10^{-12}$ molL^{-1}
pH = $-\log[H_3O_+] = 12.0$

b $KOH_{(aq)} \longrightarrow K^+_{(aq)} + OH^-_{(aq)}$
$[H_3O^+] = K_w/[OH^-] = 1 \times 10^{-14}/0.0105 = 9.52... \times 10^{-13}$ molL^{-1}
pH = $-\log[H_3O^+] = 12.0$

c $LiOH_{(aq)} \longrightarrow Li^+_{(aq)} + OH^-_{(aq)}$
$[H_3O^+] = K_w/[OH^-] = 1 \times 10^{-14}/1.06 \times 10^{-3} = 9.43... \times 10^{-12}$ molL^{-1}
pH = $-\log[H_3O^+] = 11.0$

2 a $F^-_{(aq)} + H_2O_{(l)} \rightleftharpoons HF_{(aq)} + OH^-_{(aq)}$
$K_a(HF) = \dfrac{[F^-][H_3O^+]}{[HF]}$

Assuming [HF] = $[OH^-]$ and that very little F^- has dissociated
$[OH^-] = K_w/[H_3O^+]$
$[H_3O^+] = \sqrt{K_a \times K_w/[F^-]} = \sqrt{7.2 \times 10^{-4} \times 1 \times 10^{-14}/0.01} = 2.68\ldots \times 10^{-8}$ $molL^{-1}$
$pH = -\log[H_3O^+] = 7.57$

b $NH_{3\,(aq)} + H_2O_{(l)} \rightleftharpoons NH_4^+{}_{(aq)} + OH^-_{(aq)}$
$K_a(NH_4^+) = \dfrac{[NH_3][H_3O^+]}{[NH_4^+]}$
Assuming $[NH_4^+] = [OH^-]$ and that very little NH_3 has dissociated
$[OH^-] = K_w/[H_3O^+]$
$[H_3O^+] = \sqrt{K_a \times K_w/[NH_3]} = \sqrt{5.6 \times 10^{-10} \times 1 \times 10^{-14}/0.02} = 1.67\ldots \times 10^{-11}$ $molL^{-1}$
$pH = -\log[H_3O^+] = 10.8$

c $HCO_3^-{}_{(aq)} + H_2O_{(l)} \rightleftharpoons H_2CO_{3\,(aq)} + OH^-_{(aq)}$
$K_a(H_2CO_3) = \dfrac{[HCO_3^-][H_3O^+]}{[H_2CO_3]}$
Assuming $[H_2CO_3] = [OH^-]$ and that very little HCO_3^- has dissociated
$[OH^-] = K_w/[H_3O^+]$
$[H_3O^+] = \sqrt{K_a \times K_w/[HCO_3^-]} = \sqrt{4.3 \times 10^{-7} \times 1 \times 10^{-14}/0.00150} = 1.69\ldots \times 10^{-9}$ $molL^{-1}$
$pH = -\log[H_3O^+] = 8.77$

d $HCOO^-_{(aq)} + H_2O_{(l)} \rightleftharpoons HCOOH_{(aq)} + OH^-_{(aq)}$
$K_a(HCOOH) = \dfrac{[HCOO^-][H_3O^+]}{[HCOOH]}$
Assuming [HCOOH] = $[OH^-]$ and that very little $HCOO^-$ has dissociated
$[OH^-] = K_w/[H_3O^+]$
$[H_3O^+] = \sqrt{K_a \times K_w/[HCOO^-]} = \sqrt{1.77 \times 10^{-4} \times 1 \times 10^{-14}/0.165} = 3.27\ldots \times 10^{-9}$ $molL^{-1}$
$pH = -\log[H_3O^+] = 8.48$

3 NaOH is a strong base and so it completely dissociates in water, producing more ions despite having the same concentration and leaving no NaOH species left. As the more ions present in the solution the better the conductor, it is a better conductor.
$NaOH_{(aq)} \longrightarrow Na^+_{(aq)} + OH^-_{(aq)}$
CH_3NH_2, on the other hand, is a weak base and so only partially dissociates in water, so it has less ions per unit volume and so is a weaker conductor of electricity.
$CH_3NH_{2\,(aq)} + H_2O_{(l)} \rightleftharpoons CH_3NH_3^+{}_{(aq)} + OH^-_{(aq)}$

3.10: Buffer solutions

1 Mix equal quantities of $CH_3CH_2NH_2$ and $CH_3CH_2NH_3Cl$ in 1 litre of water if you want the pH to equal the pK_a of $CH_3CH_2NH_3Cl$ or alter the ratio of each if you want a slightly different pH.
Upon addition of small amounts of HCl the base part of the buffer, in this case $CH_3CH_2NH_2$, will neutralise it by removing its proton, however the capacity of the buffer will now be altered.
$HCl + CH_3CH_2NH_2 \longrightarrow CH_3CH_2NH_3Cl$

2 **a** $[H_3O^+] = K_a \times [HF]/[NaF] = 7.2 \times 10^{-4} \times (0.1/0.05) = 0.001448$ $molL^{-1}$
$pH = -\log[H_3O^+] = 2.84$
b $pH = pK_a$ if the concentration of the acid = [conjugate base]
$pK_a = -\log K_a = -\log 1.76 \times 10^{-5} = 4.75 = pH$

3 Original buffer: $pH = pK_a$ and $[H_2C_2O_4] = [HC_2O_4^-]$
Diluted buffer: $[H_2C_2O_4] = x/3.333$ $molL^{-1}$ $= [HC_2O_4^-]$
pH is unchanged, but the buffer capacity is reduced by a factor of 3.33.

3.11: Titration curves

1 **a** A: buffer zone
B: equivalence point
b This is the buffer zone where the pK_a of the weak acid equals the pH, as it acts as a buffer as there are equal amounts of the weak acid and its conjugate base the pH does not alter upon addition of small amounts of base as the weak acid reacts with it and neutralises it.

2 **a** **Initial pH:** $[OH^-] = 0.0150$ $molL^{-1}$
$[H_3O^+] = K_w/[OH^-] = 1 \times 10^{-14}/0.0150 = 6.66\ldots \times 10^{-13}$ $molL^{-1}$
$pH = -\log[H_3O^+] = 12.2$
Equivalence point pH: The pH is 7 at equivalence point since we are reacting a strong acid with a strong base and at this point they should both be neutral.
Final pH: $HCl + NaOH \longrightarrow NaCl + H_2O$
n(HCl):n(NaOH) = 1:1
n(NaOH) = c(NaOH) x V(NaOH) = 0.0150 x 25/1000 = 3.75×10^{-4} mol = n(HCl)
c(HCl) = n(HCl)/V(HCl) = $3.75 \times 10^{-4}/(20/1000) = 0.01875$ $molL^{-1}$
c(HCl) = $[H_3O^+]$
$pH = -\log[H_3O^+] = 1.73$
b **Initial pH:** $[KOH] = [OH^-] = 0.2$ $molL^{-1}$
$[H_3O^+] = K_w/[OH^-] = 1 \times 10^{-14}/0.2 = 5 \times 10^{-14}$
$pH = -\log[H_3O^+] = 13.3$
pH at equivalence point: $CH_3COOH + KOH \longrightarrow KCH_3COO + H_2O$
$n(CH_3COOH):n(KOH) = 1:1$
n(KOH) = c(KOH) x V(KOH) = 0.2 x 15/1000 = 0.003 mol = $n(CH_3COOH)$
$c(CH_3COO^-) = 0.003/(35/1000) = 0.0857$ $molL^{-1}$
$[H_3O^+] = \sqrt{K_a \times K_w/[CH_3COO^-]} = \sqrt{1.76 \times 10^{-5} \times 1 \times 10^{-14}/0.0857} = 1.43\ldots \times 10^{-9}$ $molL^{-1}$
$pH = -\log[H_3O^+] = 8.84$
Final pH: $[H_3O^+] = \sqrt{K_a \times [CH_3COOH]} = \sqrt{1.76 \times 10^{-5} \times 0.150} = 0.00162$ $molL^{-1}$
$pH = -\log[H_3O^+] = 2.79$
c **Initial pH:** $[H_3O^+] = \sqrt{K_a \times K_w/[NH_3]} = \sqrt{5.62 \times 10^{-10} \times 1 \times 10^{-14}/0.6} = 3.06 \times 10^{-12}$ $molL^{-1}$
$K_a = \text{inv log } {-pK_a} = 5.62.. \times 10^{-10}$
$pH = -\log[H_3O^+] = 11.5$
pH at equivalence point: $[NH_4^+] = [NH_3]/(10/22) = 0.6/(10/22) = 0.273$ $molL^{-1}$
$[H_3O^+] = \sqrt{K_a \times [NH_4^+]} = \sqrt{5.62 \times 10^{-10} \times 0.273} = 1.238 \times 10^{-5}$ $molL^{-1}$
$pH = -\log[H_3O^+] = 4.91$
Final pH: $NH_3 + HCl \longrightarrow NH_4Cl$ $n(HCl):n(NH_3) = 1:1$
$n(NH_3) = c(NH_3) \times V(NH_3) = 0.6 \times 10/1000 = 0.006$ mol = n(HCl)
c(HCl) = n(HCl)/V(HCl) = 0.006/(12/1000) = 0.500 $molL^{-1}$ = $[H_3O^+]$
$pH = -\log[H_3O^+] = 0.301$

3.12: Indicators

a Bromothymol blue as its pK_a value will be the closest to the pH at equivalence point.
b Methyl red as its pK_a value will be the closest to the pH at equivalence point.
c Phenolphthalein as its pK_a value will be the closest to the pH at equivalence point.

Experiment 3: Which indicator is best?

Safety precautions

Ammonia, sodium hydroxide, hydrochloric acid and ethanoic acid are all irritants to skin and eyes they are also corrosive. Methyl red, methyl orange, phenolphthalein are skin and eye irritants as is bromothymol blue, however this is very hazardous if swallowed. Universal indicator is also an irritant, particularly if ingested or with eyes.

Method

The method should be a step-by-step guide, for example:
Step 1: Add 2 mL of hydrochloric acid to a test tube and then a few drops of methyl orange, then drop by drop add sodium hydroxide until the colour changes.
Step 2: Add 2 mL of hydrochloric acid to a test tube and then a few drops of phenolphthalein, then drop by drop add sodium hydroxide until the colour changes.
Step 3: Add 2 mL of hydrochloric acid to a test tube and then a few drops of litmus paper, then drop by drop add sodium hydroxide until the colour changes.
Step 4: Add 2 mL of hydrochloric acid to a test tube and then a few drops of red cabbage juice (crush up some red cabbage in a mortar and pestle and add a bit of hot water), then drop by drop add sodium hydroxide until the colour changes.

ISBN: 9780170352611

Step 6: Add 2 mL of hydrochloric acid to a test tube and then a few drops of methyl red, then drop by drop add sodium hydroxide until the colour changes.
Step 7: Add 2 mL of hydrochloric acid to a test tube and then a few drops of bromothymol blue, then drop by drop add sodium hydroxide until the colour changes.
If time permits, repeat these steps using HCl and NH_3 or CH_3COOH and NaOH.

Conclusion

The best indicators are the ones that change colour rapidly and within one drop this will depend upon what acids or bases you use.

Checkpoint 3: Buffers, pH and conductivity of acids and bases

1 **a** $pH = -\log[H_3O^+] = -\log 0.15 = 0.824$

b $[H_3O^+] = \sqrt{K_a \times [CH_3COOH]} = \sqrt{1.76 \times 10^{-5} \times 0.2} = 0.00187$ $molL^{-1}$
$pH = -\log[H_3O^+] = 2.73$

c $[H_3O^+] = K_w/[OH^-] = 1 \times 10^{-14}/0.608 = 1.64 \times 10^{-14}$ $molL^{-1}$
$pH = -\log[H_3O^+] = 13.8$

d $K_a(CH_3NH_3^+)$ = inv log-pK_a = inv log-10.7 = $1.99... \times 10^{-11}$
$[H_3O^+] = \sqrt{K_a \times K_w/[CH_3NH_2]} = \sqrt{1.99 \times 10^{-11} \times 1 \times 10^{-14}/0.112} = 1.33 \times 10^{-12}$ $molL^{-1}$
$pH = -\log[H_3O^+] = 11.9$

e The KOH will be a better conductor than the CH_3NH_2 as it is a strong base and so will completely dissociate in water to produce K^+ and OH^- ions; whereas CH_3NH_2 will only partially dissociate in water to produce its conjugate acid and hydroxide ions so therefore there will be less ions in solution. The more ions present in solution, the better the conductor.

$KOH_{(aq)} \longrightarrow K^+_{(aq)} + OH^-_{(aq)}$
$CH_3NH_{2\,(aq)} + H_2O_{(l)} \rightleftharpoons CH_3NH_{3\,(aq)}^+ + OH^-_{(aq)}$

2 By mixing these two solutions together you would make a buffer solution, as a buffer solution is made from a mixture of a weak acid and its conjugate base. This means that if small amounts of an acid are added, the weak base will protonate it, and if small amounts of a base are added, the weak acid will donate its proton to it and neutralise it.
$[H_3O^+] = K_a \times [CH_3COOH]/[KCH_3COO] = 1.76 \times 10^{-5} \times 0.1/0.15 = 1.17... \times 10^{-5}$ $molL^{-1}$
$pH = -\log[H_3O^+] = 4.93$

3 The initial pH would be at the start of the titration where no nitric acid has been added so it would just be the ammonia that would affect it. Since ammonia is a weak base you would calculate it using the following two formulas:
$[H_3O^+] = \sqrt{K_a \times K_w/[NH_3]}$ and $pH = -\log[H_3O^+]$
The pH at the buffer zone would be equal to the pK_a of the conjugate acid to ammonia, which is ammonium; to calculate pK_a from the K_a you would use the following formula: $pK_a = -\log K_a$
The pH at the equivalence point would be determined solely by the conjugate acid to ammonia ammonium as all the ammonia would have been reacted with the nitric acid. In order to calculate the pH of a weak acid like ammonium you would use the following formula:
$[H_3O^+] = \sqrt{K_a \times [NH_4^+]}$
The $[NH_4^+]$ would equal half the concentration of $[NH_3]$ as at this point there would be twice the volume (if the concentrations are the same, otherwise the volume will equal the aliquot + the titre). Finally you would use $pH = -\log[H_3O^+]$.
The final pH would just be the pH of nitric acid since all the ammonia will have long since been reacted. In order to calculate the pH of nitric acid you would have to write a fully balanced equation for the titration:
$NH_3 + HNO_3 \longrightarrow NH_4NO_3$
This equation shows that the moles of nitric acid are equal to the moles of ammonia; in order to calculate the moles of ammonia use its concentration multiplied by its volume. Then to calculate the concentration of nitric acid take the moles of ammonia since they are equal and divide it by the volume of nitric acid used. Since the concentration of nitric acid will be equal to the concentration of hydronium ions, you would then use $pH = -\log[H_3O^+]$.

Exam-type questions

Question one

The added acid would react with the hydroxide ions in the equilibrium removing them as shown in the following equation:
$OH^- + H_3O^+ \longrightarrow 2H_2O$
$CH_3COOH + H_2O \rightleftharpoons H_3O^+ + CH_3COO^-$
$CH_3CH(OH)COOH + H_2O \rightleftharpoons H_3O^+ + CH_3CH(OH)COOH$
This would force the equilibrium towards the reactants making less enamel and therefore weakening the coating on your teeth.
When you brush your teeth, toothpaste, which is made of a weak base, adds hydroxide ions into the your mouth moving the equilibrium towards the enamel and decreasing its solubility.

Question two

a $K_s = [Be^{2+}][OH^-]^2 = 4s^3$
$s = \sqrt[3]{K_s/4} = \sqrt[3]{6.92 \times 10^{-22}/4} = 5.57 \times 10^{-8}$ $molL^{-1}$

b $Be(OH)_{2\,(s)} \rightleftharpoons Be^{2+}_{(aq)} + 2OH^-_{(aq)}$
If the pH was raised there would be more hydroxide ions in solution therefore the beryllium hydroxide would become less soluble as the equilibrium would shift in favour of the reactants in order to correct the change made to the system, according to Le Chatelier's principle.

Question three

a $Zn(OH)_{2\,(s)} \rightleftharpoons Zn^{2+}_{(aq)} + 2OH^-_{(aq)}$
By raising the pH there will be more hydroxide ions present in the solution; this will force the equilibrium to shift towards the solid which will decrease the solubility and form more zinc hydroxide.
$[H_3O^+]$ = inv log -pH = inv log -9.2 = $6.30... \times 10^{-10}$ $molL^{-1}$
$[OH^-] = K_w/[H_3O^+] = 1.58 \times 10^{-5}$ $molL^{-1}$
$K_s = [Zn^{2+}][OH^-]^2$
$K_s = s \times (s + 1.58 \times 10^{-5})^2$
We assume s is much less that 1.58×10^{-5} $molL^{-1}$
$s = K_s/1.58 \times 10^{-5} = 3 \times 10^{-17}/1.58 \times 10^{-5} = 1.89 \times 10^{-12}$ $molL^{-1}$
Since s is much less than the hydroxide ion concentration there will be more precipitate formed.

b By lowering the pH there will be less hydroxide ions present in solution and so the zinc hydroxide will become less soluble as there will be less hydroxide ions available in solution and so the equilibrium will shift towards the ions.

c The ammonia will form a complex ion with the zinc $[Zn(NH_3)_4]^{2+}$, which will shift the equilibrium towards the ions and so therefore decrease the solubility of the zinc hydroxide.

Question four

a In order for a precipitate to form, $K_s = Q_s$, so $Q_s = [Ag^+][Cl^-] = 0.1 \times [Cl^-]$
$[Cl^-] = K_s/[Ag^+] = 2 \times 10^{-10}/0.1 = 2 \times 10^{-9}$ $molL^{-1}$
$n(Cl^-) = c \times V = 2 \times 10^{-9} \times 1 = 2 \times 10^{-9}$ mol
$m(Cl^-) = M \times n = 74.6 \times 2 \times 10^{-9} = 1.49 \times 10^{-7}$ g (3 s.f.)

b Silver will form a complex ion with the ammonia molecules which will decrease the solubility of the silver chloride since there will be less silver ions present in the equilibrium. The complex ion is $Ag(NH_3)_2^+$ and the original equilibrium is
$AgCl_{(s)} \rightleftharpoons Ag^+_{(aq)} + Cl^-_{(aq)}$

Question five

a K_a = inv log -pK_a = inv log -6.91 = 1.23×10^{-7}
$[H_3O^+] = \sqrt{K_a \times [HSO_4^-]} = \sqrt{1.23 \times 10^{-7} \times 0.501} = 2.48 \times 10^{-4}$ $molL^{-1}$
$pH = -\log[H_3O^+] = 3.61$

b $[H_3O^+] = \sqrt{K_a \times K_w/[SO_4^{2-}]} = \sqrt{1.23 \times 10^{-7} \times 1 \times 10^{-14}/0.166} = 8.60... \times 10^{-11}$ $molL^{-1}$
$pH = -\log[H_3O^+] = 10.1$

Question six

$CH_3CH_2NH_2$ is a weak base and so it will partially dissociate in water to produce its conjugate acid and hydroxide ions as shown in the following equation:
$CH_3CH_2NH_2 + H_2O \rightleftharpoons CH_3CH_2NH_3^+ + OH^-$
As it only partially dissociates the highest concentration of any

species present in its solution will be itself, $CH_3CH_2NH_2$. Then comes the products of its dissociation, $CH_3CH_2NH_3^+ + OH^-$, which will be present in relatively equal concentrations. Finally there will be a small amount of H_3O^+ ions as water always partially dissociates into OH^- and H_3O^+ but there will be much less H_3O^+ than anything else.

KHCOO will dissolve in water to produce K^+ and $HCOO^-$ ions as shown in the following equation: $KHCOO_{(aq)} \longrightarrow K^+_{(aq)} + HCOO^-_{(aq)}$

The $HCOO^-$ ions are a weak base and so it will partially dissociate further as shown in the following equation:

$HCOO^- + H_2O \rightleftharpoons HCOOH + OH^-$

So K^+ will be the highest concentration as it does not react further with water, then comes $HCOO^-$ as it only partially dissociates, further followed by the products of its dissociation, HCOOH and OH^-, which will be relatively equal in concentration.

Finally there will be a small amount of H_3O^+ ions as water always partially dissociates into OH- and H_3O^+ but there will be much less H_3O^+ than anything else.

Question seven

KCl has no acidic or basic ions making it up hence it has a neutral pH. It is a good conductor of electricity as it dissolves in water completely producing K^+ and Cl^- ions.

$KCl_{(aq)} \longrightarrow K^+_{(aq)} + Cl^-_{(aq)}$

The more ions present in solution, the better the conductor.

CH_3CH_2COOH has an acidic pH as it is a weak acid and so it dissociates partially in water to produce the acid particle, H_3O^+:

$CH_3CH_2COOH + H_2O \rightleftharpoons CH_3CH_2COO^- + H_3O^+$

It is a weak conductor of electricity as it only partially dissociates and so therefore it doesn't have a high concentration of ions in solution.

CH_3CH_2COONa will completely dissolve in water to produce $CH_3CH_2COO^-$ and Na^+ ions, therefore there will be a high concentration of ions present in solution available to conduct a current.

$CH_3CH_2COO^-$ will then react as a weak base and partially dissociate further as well as protonate water. This makes the pH higher than 7.

$CH_3CH_2COO^- + H_2O \rightleftharpoons CH_3CH_2COOH + OH^-$

Question eight

a Since the concentrations of NH_3 and NH_4Cl are the same, the pK_a equals the pH of the buffer.

$pK_a = -\log K_a = -\log 5.6 \times 10^{-10} = 9.25$

$pH = pK_a = 9.25$

b Upon addition of small amounts of hydrochloric acid the ammonia will protonate it and therefore neutralise it without altering the pH of the buffer. However it will reduce the capacity of the buffer.

$HCl + NH_3 \longrightarrow NH_4Cl$

Question nine

Initial pH: $[H_3O^+] = \sqrt{K_a \times [CH_3COOH]} = \sqrt{1.76 \times 10^{-5} \times 0.500} = 0.00296$ $molL^{-1}$

$pH = -\log[H_3O^+] = 2.53$

Buffer zone pH: $pK_a = -\log K_a = -\log 1.76 \times 10^{-5} = 4.75 = pH$

Equivalence point pH: $[CH_3COO^-] = 0.50/2 = 0.25$ $molL^{-1}$

$[H_3O^+] = \sqrt{K_a \times K_w/[CH_3COO^-]} = \sqrt{1.76 \times 10^{-5}\ 1 \times 10^{-14}/0.25}$

$= 8.39 \times 10^{-10}$ $molL^{-1}$

$pH = -\log[H_3O^+] = 9.08$

Final pH: $c(NaOH) = [OH^-] = 0.500$ $molL^{-1}$

$[H_3O^+] = K_w/[OH^-] = 1 \times 10^{-14}/0.5 = 2 \times 10^{-14}$ $molL^{-1}$

$pH = -\log[H_3O^+] = 13.7$

Question ten

Initial pH: $[H_3O^+] = \sqrt{K_a \times K_w/[NaC_6H_5COO]} = \sqrt{6.46 \times 10^{-5} \times 1 \times 10^{-14}/0.0150} = 6.56 \times 10^{-9}$ $molL^{-1}$

$pH = -\log[H_3O^+] = 8.18$

Buffer zone pH: $pK_a = -\log K_a = -\log 6.46 \times 10^{-5} = 4.19 = pH$

Equivalence point pH: $[C_6H_5COOH] = 0.0150/2 = 0.00750$ $molL^{-1}$

$[H_3O^+] = \sqrt{K_a \times [C_6H_5COOH]} = \sqrt{6.46 \times 10^{-5} \times 0.0075} = 6.96 \times 10^{-4}$ $molL^{-1}$

$pH = -\log[H_3O^+] = 3.16$

Final pH: $HCl + NaC_6H_5COO \longrightarrow NaCl + C_6H_5COOH$

$n(HCl): n(NaC_6H_5COO) = 1:1$, so $n(NaC_6H_5COO) = n(HCl) = c(NaC_6H_5COO) \times V(NaC_6H_5COO) = 0.015 \times 15/1000 = 2.25 \times 10^{-4}$ mol

$c(HCl) = n(HCl)/V(HCl) = 2.25 \times 10^{-4}/(15/1000) = 0.0150$ $molL^{-1}$

 ISBN: 9780170352611